D0311031

IN SEARCH OF TIME

IN SEARCH OF TIME

THE HISTORY, PHYSICS, AND PHILOSOPHY OF TIME

DAN FALK

THOMAS DUNNE BOOKS
ST. MARTIN'S GRIFFIN
NEW YORK

THOMAS DUNNE BOOKS.
An imprint of St. Martin's Press.

IN SEARCH OF TIME. Copyright © 2008 by Dan Falk. All rights reserved.
Printed in the United States of America. For information, address St.
Martin's Press, 175 Fifth Avenue, New York, N.Y. 10010.

www.thomasdunnebooks.com
www.stmartins.com

The Library of Congress has cataloged the hardcover edition as follows:

Falk, Dan.
    In search of time : the science of a curious dimension /
Dan Falk.—1st ed.
      p. cm.
    Includes bibliographical references and index.
    ISBN 978-0-312-37478-5
    1. Time—History. 2. Science and civilization. I. Title.
QB209.F35 2009
529—dc22

2008024875

ISBN 978-0-312-60351-9 (trade paperback)

First published in Canada as *In Search of Time: Journeys
Along a Curious Dimension* by McClelland & Stewart Ltd

First St. Martin's Griffin Edition: January 2010

10 9 8 7 6 5 4 3 2 1

In memory of my grandparents
Ignacy and Leonia Falk
Moshe Raviv and Dr. Rosalie Shein

That great mystery of TIME, were there
no other; the illimitable, silent,
never-resting thing called Time, rolling,
rushing on, swift, silent, like an
all-embracing ocean tide, on which we
and all the Universe swim like exhalations,
like apparitions which *are*, and then
are not: this is forever very literally
a miracle; a thing to strike us dumb,
– for we have no word to speak about it.

– Thomas Carlyle, Heroes and Hero Worship (1840)

CONTENTS

PREFACE

"You're writing a book about *what*?"

Tell people you're working on a book about time, and you get some interesting reactions. Some look confused, or give a dismissive shrug – "What *about* time," they ask, as though there could hardly be enough interesting things to say about it to fill an entire book. ("Doesn't it just sort of tick by?") Others seem to understand the appeal right away, and wonder about particular topics. "Will you talk about time travel?" Yes, I affirm – it gets a whole chapter to itself. (Even if time travel is impossible, I tell them, it raises fascinating questions about the nature of time, space, and the laws of nature.) Some people guess that I must be writing a "physics book" – highly technical, with much talk of entropy and world lines and such. Not so, I assure them. Or at least not "just" a physics book. My goal is to take a broader approach, tackling the enigma of time from several different directions, each with its own perspective and insights – and its own track record of successes and frustrations.

In fact, one *has* to approach time from many such angles, because no one discipline has "the answer." That fact was apparent whenever I glanced at the books that line the shelves in my apartment. (There would, of course, be many trips to many libraries, but one of the joys of building up a decent home collection is that quite a good chunk of research can be done before one even braves the elements.) My first couple of shelves cover the history and philosophy of science: this is where I reach for classics like Bronowski and Boorstin and Gamow, a cluster of Carl Sagan titles, and more recent works by Timothy Ferris and Dennis Danielson, to name a few. Beneath that, the scientific biographies: Drake and Sobel on Galileo; Westfall and Gleick on Newton; Pais, Fölsing, and Isaacson on Einstein; and other more narrowly focused titles on the theories crafted

by those great minds. Under that, a shelf of modern physics and cosmology titles: Hawking, Weinberg, Greene, Davies, Rees, Krauss, and others.
On the next shelf, books on evolution and the nature of the human
animal: Diamond, Tattersall, Dawkins, Hauser. Below that, books on consciousness and the mind: Pinker, Penrose, Dennett, Crick, Damasio,
Edelman. Not to mention some very thorough books on clocks, calendars, and timekeeping: Whitrow, Aveni, Landes, Duncan, Steel.

And *time* – well, time reaches across *all* of those fields. In fact,
one of the challenges is that each of these disciplines is to some degree
intertwined with each of the others. Interconnectedness is fine if you're
building a spider web, but it can actually hamper the writing of a book,
which demands a single flowing narrative – a story. To tell that story, I've
had to be selective. Where there was a choice between more science and
more philosophy, the science usually won out – not because the philosophy is uninteresting, but merely because I found it to have less power
to advance the story. ("What, no Heidegger? No Bergson?" Sadly, no;
we will have to make do with Plato, Aristotle, Leibniz, McTaggart, and a
handful of other key players.) Even within the sciences, there is far too
much current research to assess in a single volume; indeed, each of my
twelve chapters could easily have been a book in its own right, for an
author so inclined. For those who do wish to probe deeper, I hope that
detailed notes and a full bibliography of sources will aid in further
reading. In the text itself, I have tried to make my selections as carefully
as possible, giving the most attention to those areas where science has
made the most tangible advances in recent years.

Research begins with books and journals and libraries, but it does
not end there: over the last few years, it has been my great privilege to
meet with some of the deepest thinkers of our time, in some cases for
multiple interviews. "It will only take an hour," I would tell them, knowing
full well that it would run much longer – and most of them generously
allowed me to keep asking away, microphone in hand. I am especially
grateful to Roger Penrose, Julian Barbour, David Deutsch, Lee Smolin,
and Paul Davies; their grasp of some of the most difficult problems in all

of science is truly inspiring. Many more scholars sat down with me and patiently described their research; others guided me through laboratories, museum exhibits, and archeological sites. They are named in the chapters ahead, and I am indebted to them all. (Most of these interviews were conducted specifically for the present book, but I have also occasionally drawn on research for various earlier projects, including several documentaries that I made for the CBC Radio program *Ideas*.)

A number of people generously looked over portions of the manuscript at various stages; Ivan Semeniuk, George Musser, and Natalie Munro made especially valuable comments, while Elizabeth Howell was kind enough to read the entire text. (Of course, any errors that remain are my responsibility alone.) I also benefited greatly from discussions with James Robert Brown, Glenn Starkman, and Eugenie Scott.

The idea of writing a book about time had been in the back of my mind since the completion of my first book, *Universe on a T-Shirt: The Quest for the Theory of Everything* (2002). This book is by no means a sequel; the subject matter is, in general, quite different. But certain key topics – special relativity, for example – do come up again; on occasion, I refer the reader to more detailed explanations in the earlier book.

This project would not have come to fruition without the help of my agents, Don Sedgwick and Shaun Bradley of the Transatlantic Literary Agency, and especially the tireless work of my editor, Jenny Bradshaw at McClelland & Stewart, who helped hone the manuscript into its finished form. I am also grateful to Stephanie Fysh for her copy editing skills.

Where measurements crop up, I have once again primarily used metric units; I trust that my U.S. readers will not have too much trouble with meters, kilometers, and the like. On the other hand, I have used American spellings, which I hope will not alienate my Canadian audience.

Readers' comments are welcome; I can be reached through my website at www.danfalk.ca.

IN SEARCH OF TIME

INTRODUCTION

If we are aware of anything we are aware of the
passage of time.

— J.R. Lucas, *A Treatise on Time and Space*

Time passes. Listen. Time passes.

— Dylan Thomas, "Under Milk Wood"

" 've completely solved the problem," a young Albert Einstein said
excitedly to his friend Michele Besso in May of 1905. "My solution
was to analyze the concept of time."

Besso, a colleague of Einstein's at the patent office in Bern,
Switzerland, was the first to be let in on the secret. A month later, the whole
world would know. (Or at least, those who read the *Annalen der Physik* on
a regular basis. Another fourteen years would pass before the scientist
would become a household name.) Einstein's groundbreaking paper, the
result of ten years of intense study and ingenious "thought experiments,"
was an attempt to reconcile Maxwell's theory of electromagnetism with
long-established ideas about relative motion dating back to Galileo. It was
an urgent problem that had stumped the brightest minds of the day. The
paper appeared under the innocuous title "On the Electrodynamics of
Moving Bodies" – and it changed everything. Suddenly time was flexible,
like rubber. Space and time were intimately linked. And simple words like
"now" seemed to lose their meaning entirely.

Einstein's paper was shocking precisely because time is – or was supposed to be – so simple. It still *seems* simple, more than a hundred years after Einstein's breakthrough. Time, after all, surrounds us. It envelops and defines our world; it echoes through our every waking hour. Time is the very foundation of conscious experience.

Above all, time flows – or appears to flow. A river is the favorite metaphor; we imagine time as a relentless stream, bringing the future toward us and carrying past events off behind us. Equivalently, we can imagine time as a fixed landscape, with ourselves sailing through it. A more modern metaphor is the movie projector: events can be likened to frames of film, each illuminated ever so briefly in the light of an instantaneous "now," after which they recede into the past. Future events – later frames – rush toward the lens, each to experience its own brief "now" in due course.

By either metaphor, time appears to flow in just one direction, leading from the fixed events of the past toward an unknown future. It is unrelenting. No sooner have we uttered the word "now" than another "now" has taken its place. That previous "now" is lost to the past, gone forever. We cannot change the events of five seconds ago any more than we can revisit the Battle of Hastings. The future, for its part, is hurtling toward us, unstoppable. We may be unsure of what it will bring – but of its arrival there can be no doubt.

If these statements seem obvious, perhaps even juvenile, it is a reflection of just how deeply ingrained such feelings are. Young children quickly learn the meaning of words like "yesterday," "today," and "tomorrow"; of "past," "present," and "future." We think of time as a commodity: we try to *save time*; we hate to *waste time*; we say we'll *make time* for some favorite activity. When we need to catch our breath, we call a *time-out*. We say that time flies when we're having fun and slows to a crawl at the dentist's office – but deep down we know better. We trust the keeping of "true" time to our clocks, which, in a world peppered with semiconductor-laden gadgets, are more ubiquitous than ever. Yet we also suspect that time would go by just as relentlessly even if no clocks were around to mark its passing. As Aristotle remarked some 2,300 years ago, "Even when it is dark, and we are not being affected through the body, if any process goes on in the mind, immediately we think that some time has elapsed too." Isaac Newton suspected that time would flow just the same even if *nothing* were around to mark its passing – but, as we will see, Newton does not have the last word on such matters. Nor, for that matter, does Einstein: the problem that he "completely solved" in 1905 was just one of time's many mysteries. Time has not yet given up all its secrets.

The great paradox of time is that it is at once intimately familiar and yet deeply mysterious: nothing is more central and yet so remote. To be human is to be aware of the passage of time; no concept lies closer to the core of our consciousness. Yet who can say just what time is? It is thoroughly intangible. We cannot see, hear, smell, taste, or touch it. Yet we do *feel* it. Or at least we *think* we feel it. This, as we shall see, is not just wordplay: scientists and philosophers continue to debate just what we mean when we utter a simple phrase like "time passes."

Time is linked to change. At one time, we observe *this*; at a later time, we observe *that*; and we associate that change with time's passing. No wonder that time is sometimes defined as "nature's way of keeping everything from happening at once." And yet to equate time with change seems to be missing something. The passage of time seems, somehow, to be more fundamental. No wonder poets, writers, philosophers, and scientists have grappled with the idea of time for centuries.

So again we ask: What is time? A child might say, "It's that stuff that flows by, even when you're standing still," or, perhaps, "It's what clocks measure." Can a grown-up say much more? "It's a dimension, like space," remarks someone with a rudimentary idea of Einstein's breakthrough – and yet it feels very different from space.

The problem with these (and many other) statements about time is that they often seem less and less satisfying the more closely we examine them. We said that time "surrounds us" and "defines our world" – but is that true for everyone, or is it something associated primarily with time-obsessed Western culture? Would a Buddhist monk be as worried about an appointment as a Wall Street broker? We observe that children learn to say "past," "present," and "future" – but only in cultures where parents consider those terms important. In some cultures, as we'll see, those words – perhaps those concepts – are missing.

And then there is that most basic feeling – the feeling that time "flows." But what meaning can we attach to such a statement? We say time flows like a river . . . but a river flows with respect to the shore. Time flows with respect to . . . what? A river flows at the rate of, say, a thousand gallons per second; time flows by at a rate of . . . one second per second? Such a declaration tells us nothing. (In fact, such an assertion would force us to imagine a secondary time or "hypertime" against which our primary time flows. And if *that* time flowed, we'd need a third time, and so on. Not a very helpful state of affairs!) No wonder St. Augustine of Hippo (A.D. 354–430), who spent years pondering the problem of time, had moments of deep frustration. "What, then, is time? If no one ask of me, I know," he laments. "But if I wish to explain to him who asks, I know not." Augustine eventually came to suspect that time is all in our heads, merely something we construct with our minds. Other philosophers, over the centuries, have come to the same conclusion. Yet time feels more real than that, doesn't it?

Science has helped – but it has also deepened the mystery. Einstein's relativity showed us that such everyday concepts as "now" lose their meaning in four-dimensional spacetime. What time is it "now" in the Andromeda Galaxy? There is no meaningful answer. Don't feel bad if you're troubled by this predicament; it troubled Einstein, too, as we will see.

More shocking still, physics makes no distinction between past and future. Some physicists think of time, together with space, as a vast block in which past and future have equal status. "Now," meanwhile, is reduced to a subjective label, just like "here." Some scientists feel that, while time itself may be real, its flow or passage is purely illusory, an artifact of the way conscious observers perceive their surroundings. No conscious observer, no passage of time. Echoes of Augustine.

Our struggle to comprehend what time is has never gotten in the way of our obsession with measuring it. While philosophers and physicists continue to grapple with its meaning, craftsmen and tinkerers from all cultures have shown boundless creativity in tracking time's passage on both the smallest and the largest scales.

Human beings have probably been tracking time in one way or another for as long as our species has existed. The most obvious natural cycles – the day, the lunar month, the year – would have commanded the attention of our ancestors. (Unlike today's city-dwellers, they would have enjoyed pitch-black skies at night. Their lives were inevitably touched by the comings and goings of heavenly bodies.) Signs of ritual burial, including grave goods that hint at a conception of eternity, go back tens of thousands of years.

In historical times, the record is far clearer. Every ancient civilization developed calendars to track nature's cycles – in many cases, with remarkable sophistication. Our own calendar traces its roots to Egypt and Babylon, with a few modern revisions. By inserting a leap year every four years (thank you, Julius Caesar) and *not* inserting a leap year three times every four hundred years (thank you, Pope Gregory XIII), we managed to cobble together our days into years in a way that mimics, with reasonable accuracy, nature's cycles. As we'll see, however, the Gregorian calendar is just one of many solutions to the problem of keeping pace with those cycles.

In some ancient cultures, time was seen as cyclic, with sequences of events repeating over and over again. For some cultures, death itself is seen merely as a transition to a new state of existence, human or otherwise. Judeo-Christian theology imagines life beyond death, but its view of history is quite different: events unfold in a unique sequence under God's watchful eye, from a singular moment of Creation to an eventual Day of Judgment – a decidedly *linear* view of time. The idea of linear time, historians have argued, became a cornerstone of the Western world view. It may have paved the way for the Scientific Revolution and the Industrial Revolution, which in turn triggered an affinity for reason and a sense of progress. By the end of the seventeenth century, Europeans viewed time as an abstract entity, something wholly independent of human activity.

Time is now everywhere: the seconds tick away on digital watches, cell phones, and computers; the electronic networks that connect our world rely on signals from atomic clocks synchronized to billionths of a second. In the Olympics, a hundredth of a second could mean the difference between a gold medal and a silver – but that blink-of-an-eye interval is an eternity compared to the shortest times measured by physicists, who can discern happenings as brief as 100 *attoseconds*. (How short is that? One hundred attoseconds is as small compared to one second as one second is to 300 million years.)

Human beings pay more attention to time than any other species – but all living creatures respond to time's cycles. Plants and animals have internal clocks that keep their rhythms in tune with their natural environment. The organ most directly responsible for our perception of time is, of course, the brain. Somehow we take in a vast array of chaotic sensory data from our environment and organize it into a meaningful picture of our surroundings. But it is an ever-changing picture; it is a picture that evolves in time, a picture rooted in time. Human beings have a remarkably sophisticated ability to form, store, and recall these mental images. Memory, it seems, is *all about* time. "Now" may last just a flickering moment, but in our minds it can endure for decades. A memory of an especially powerful experience – a first kiss, the birth of a child, the death of a loved one – can last a lifetime.

We not only remember the past, we also envision the future. In fact, we can project our minds across the ages. With relative ease we can picture a Roman centurion or an interstellar spacecraft. The images may not be accurate – they may be mere caricatures – but the fact that we can even entertain such ideas sets us apart from all other living things. We are creatures of time, embedded in time.

Even if there were no historians or archeologists – no human beings, even – the universe itself would still record its own past. It is not always easy to decipher these records, but with the right tools we can read nature's own history books. Fossils, for example, tell of ancient plant and animal species, many of them long extinct. Radioactive atoms can tell us how long ago they thrived. A canyon tells of millennia of weathering and erosion. The cosmos itself, as astronomers have found, carries echoes of its own youth – photons of light that have danced across the cosmos for nearly 14 billion years.

That mind-numbing figure – 14 billion years – is our best guess at the age of the universe, our best estimate of just how much time has passed so far. In our final chapters, we'll examine the evidence for this remarkable finding. We will also look ahead, and speculate on how much time remains. In all likelihood, there is much more in front of us than behind us; the universe appears to be young. But even that span, from the universe's fiery beginnings until now, is staggering. It is far longer than the time span during which ape-like creatures have walked upright on our planet, which in turn dwarfs the amount of time we have spent crafting calendars and clocks, and using the tools of science to probe our world. In the last few decades, educators have made a concerted effort to simulate that vast stretch of time visually – for example, by marking off the eons along a vast curving walkway (as at the Rose Center for Earth and Space in New York) or with a giant yellow tape measure (as at the Ontario Science Centre in Toronto), or with a fossil-laden nature trail showcasing Earth's geological history (as at the new "Trail of Time" at the Grand Canyon). Such representations, in effect, trade time for space: we can't see time, but we can see its more tangible reflection in wood or fiberglass or steel. Perhaps that is the best we can do as we struggle to picture time.

During those billions of years, who knows how many species on how many millions of planets may have evolved. For all we know, some of those creatures may have contemplated the nature of time. That, of course, is speculation. What we do know is that at least one species – *Homo sapiens* – has done so. Indeed, we have become rather obsessed with this most curious dimension.

In the chapters ahead, we'll examine what some of our most insightful thinkers have had to say about time, from Aristotle to Newton to that young man from the patent office, Albert Einstein. We will also meet many of today's most profound thinkers: Roger Penrose, Paul Davies,

Julian Barbour, David Deutsch, Lee Smolin, and many more. We'll look at what philosophers, physicists, psychologists, and neuroscientists have discovered about time, and what different cultures – today and in centuries past – have had to say about its elusive nature and its apparent flow. It will not be anything close to a complete investigation, if such a thing can even be imagined. It will, instead, be a brief tour – but, I hope, a stimulating one.

HEAVENLY CLOCKWORK

Time's natural cycles

The first grand discovery was time, the landscape of experience.

– Daniel Boorstin, *The Discoverers*

The city of Drogheda, just half an hour north of Dublin by Irish Rail's InterCity service, isn't at the top of most tourists' Ireland itineraries. Even *Lonely Planet*, which praises the surrounding counties for their historical and cultural wealth, describes this small coastal city as "charmless." As my taxi heads west, however, the view steadily improves, with Drogheda's industrial clutter giving way to the rolling hills and green valleys of County Meath. And just a few kilometers farther inland lies one of the most important prehistoric monuments in all of Europe – the "passage tomb" of Newgrange.

Most visitors to Newgrange approach from the south side of the Boyne, through the main visitor center, but for my early morning appointment I have to come from the north side of the famous river, past Newgrange Farm, where the sound of birds and cowbells fills the morning air. As the taxi rounds the last bend in the road, the monument itself – a shallow, circular, grass-covered mound some eighty meters across and a dozen meters high – comes into view. The outer walls are lined with blocks of white quartz that glitter in the sunlight. I'm met by Claire Tuffy of the Office of Public Works, which manages the site, and together we climb the gentle hill leading to the tomb's main entrance.

The structure, Tuffy explains, dates from around 3100 B.C. – making it five centuries older than the Great Pyramid at Giza in Egypt, and a full thousand years older than the "trilithons" at the center of Stonehenge. The Neolithic people who lived in Ireland at that time would have been farmers, tending grain crops and herding livestock. The Boyne – Tuffy nods toward the river, half-hidden by trees and low hills – was their highway. They had likely been farming the land for a thousand years before construction at Newgrange began. "Their tools were stone and wood – no metal," Tuffy points out. The quartz was transported from what is now County Wicklow, some eighty kilometers away. One can only imagine the Herculean effort required to move, shape, and lift the nearly two thousand stones used to construct the monument.

We pass the richly decorated sandstone blocks that mark the entrance, and approach the iron gate that now protects the interior. Tuffy unlocks it and we step inside, ducking our heads because of the low ceiling. Though circular on the outside, the interior of the tomb is long and narrow, cutting deep into the center of the mound. We make our way cautiously toward the rear of the chamber. Soon the entrance is little more than a tiny square of light in the distance behind us. Without the electric lights installed overhead every few meters, it would be pitch black. It is also eerily quiet. For the pair of bats that have built their nest here, it is no doubt the perfect home.

The tomb stretches twenty-five meters in length but rarely spans more than a meter across. At the far end are three small alcoves that branch off from the main passage, giving the tomb an elongated cruciform design. Though known to the modern Irish since the seventeenth century, the site wasn't properly excavated until the 1960s, when archeologist Michael O'Kelly and his team discovered the cremated bones of at least five individuals, on basin-like stones in the alcoves at the rear of the tomb. Workers also uncovered intriguing Neolithic artwork. Several of the stones are decorated with geometrical patterns; the most complex is a trio of overlapping spirals at the very back of the tomb.

As we stand in the rear chamber, Tuffy shines her flashlight at the layers of cobbled stones that arch overhead. "That roof has not been restored, and it is still perfectly watertight after five thousand years of Irish weather," she says. Why would Neolithic farmers have gone to that much trouble to protect the bones of the dead from a little water? Perhaps the spirits of the ancestors were thought to live on, she speculates. And while today is sunny, Tuffy reminds me with a smile that her country has been known to have the occasional spot of rain. "Maybe it's every Irish person's idea of heaven, to be dry for ever and ever," she muses.

But the most intriguing feature of Newgrange is neither its walls, nor its roof, nor its artwork. It is not what one sees at any one place, but rather what one sees at a particular *time*. Every winter, on the morning of the shortest day of the year – the winter solstice – sunlight streams through a small opening above the main entrance, known as the "roof box," illuminating the back of the tomb. This seemingly innocuous event – a sliver of sunlight briefly creeping into a dusky burial chamber in the dead of winter – is what makes Newgrange unique. These weathered stones allow us to glimpse, however dimly, into the minds of those who first considered the matter of time.

The Sun in the Cave

Were the roof box and passage constructed along a slightly different angle, the solstice event would not happen. Could it be a coincidence, a chance alignment? Almost certainly not, says astronomer Tom Ray of the Dublin Institute for Advanced Studies, who investigated the geometry of Newgrange in the 1980s. "As an astronomer, as a mathematician, I'm looking at the statistics and saying, the chances of this being an accident are very, very small," Ray told me when I visited him in his Dublin office. The solar alignment "was their intention." Andrew Powell, writing in a prominent archeological journal a few years ago, reached the same conclusion: "There can be no doubt that this was an integral feature of the tomb's design."

The Neolithic "passage tomb" at Newgrange, in Ireland, dates from 3100 B.C. On the morning of the winter solstice, sunlight passes through an opening above the tomb's entrance, illuminating the rear of the chamber. *(Above: University of Notre Dame)*

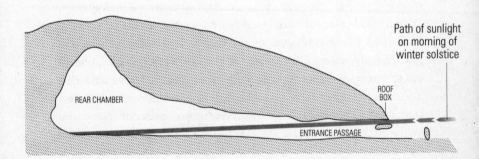

Path of sunlight on morning of winter solstice

ROOF BOX

REAR CHAMBER

ENTRANCE PASSAGE

The modern name Newgrange comes from the Gaelic *Uaimh na Gréine*, "The Cave of the Sun." In fact, even before those first excavations in the Sixties, there were local legends about sunlight entering the cave at a particular time of year. O'Kelly wondered if it might involve the winter solstice, as such alignments are well known in later Neolithic monuments. So he camped out in the cave overnight, waking up early on the morning of December 21, 1969, to see for himself (and, just to be sure, he repeated the experiment the following year). Ray, telling me the story, is mildly amused. "From the point of view of an astronomer, you don't need to actually do that," he says. "You could have just done a bit of surveying, and that would have told you the answer. So we have this sort of romantic image of Michael O'Kelly stuck in the back of the chamber, on the shortest day of the year, waiting for the sun to rise. And lo and behold, when it did, he realized that the light came into the main chamber of Newgrange." As O'Kelly noted in his journal that morning:

> At exactly 8:54 hours GMT the top edge of the ball of the sun appeared above the local horizon and at 8:58 hours, the first pencil of direct sunlight shone through the roof-box and along the passage to reach the tomb chamber floor as far as the front edge of the basin stone in the end recess.

But our view today is not precisely the same as it would have been five thousand years ago. For one thing, Ray explains, the earth's axis wobbles periodically over thousands of years. The phenomenon is related to the precession of the earth's rotation axis: as the earth spins, the axis itself gradually revolves with respect to the solar system, in a cycle that lasts about 26,000 years. The related wobble – astronomers call it "nutation" – causes a periodic change in the tilt of the axis. Today, the axis is tilted 23.5 degrees; at the time Newgrange was built, the tilt was slightly greater – about 24 degrees. The upshot of that small shift, Ray explains, is that the shortest day of the year was a little bit shorter back then, and the longest day was a little bit longer. That, in turn, affects the time of sunrise and sunset. Today, on the solstice – as O'Kelly noticed – several minutes pass between sunrise and the first penetration of sunlight into the rear of the tomb. Five thousand years ago, however, it would have been bang on. "You would have captured the sun just as it rose," says Ray.

The solstice sunrise at Newgrange is still a remarkable event. Every year, thousands of people take part in a lottery for the privilege of a morning visit to the tomb in the week of December 21. Because the main passageway is angled upward, those first rays of sunlight don't hit the back wall; instead, as O'Kelly observed, they hit the ground a few meters short of the rear of the tomb. It is that first strike of sunlight inside the tomb that those lucky visitors watch for. "Nobody takes their eye off the ground," says Tuffy, who has probably witnessed the event more often than anyone else in recent years. "You lose your sense of how much time has passed. And despite the fact that everybody is watching the floor, people inevitably miss the first beam of light." Soon the beam is about the length and width of a pencil, Tuffy says. "And then very quickly it gets longer and wider, and it moves down the floor." By the time it reaches the middle of the chamber – just a few minutes later – it's nearly twenty centimeters wide, and surprisingly intense. "It's a lovely, warm color," she says. "And then the whole room is so bright you can see right up as far as the capstone. You can see the faces of all the people gathered in the room."

What does the solar alignment tell us about the Neolithic people who moved so many massive stones into place, in such a precise way, to build such a structure? We can imagine that the builders of Newgrange – as one could say of people in any farming community today – took an intense interest in the passing of the seasons and the motions of the heavenly bodies, particularly the sun and moon. In those days, before the glare of streetlights and shopping malls, the skies would have been pristine. True, Irish weather is often rather gray – but on every clear, cloudless night, the sky would have presented a dynamic celestial display for all to see. The regularity of the heavens – the daily rising and setting of the sun, the monthly waxing and waning of the moon, the annual parade of the seasons – would have been impossible to ignore.

"Certainly they had an interest in astronomy," says Ray. He cautions, though, against projecting modern Western terms onto a culture so different from ours; we have to be careful about labeling those Neolithic farmers "astronomers" or calling Newgrange an "observatory" – although, as we'll see, scholars of all stripes can hardly refrain from using that particular term (one is tempted to call it "the 'o' word") when discussing elaborate Neolithic sites. Still, there is no question that those early farmers had a keen awareness of the heavens. "There clearly was an interest in the two main heavenly bodies," says Ray. "Whether there were sort of religious connotations as well, I don't know, and I don't think anybody else knows either."

The First Hominids

Our earliest ancestors had no clocks and calendars, but they had something that functioned in a similar way: nature itself. Early humans must have been captivated by time's endless cycles, as reflected in the rhythmic motions of the heavenly bodies, for many thousands of years. Today we look down at our watches (and the LCD clocks on our cell phones); our ancestors would have looked up at the sun, moon, and stars. And very likely a more basic awareness originated much earlier, from the time our ancestors first walked upright and chipped away at the first crude stone tools. But extrapolating from bones and tools to thoughts and beliefs is an enormous and endlessly frustrating challenge, and even the most plausible ideas are rarely proven beyond all doubt. Such efforts have been bolstered in recent years by remarkable advances in genetics, cognitive science, primate studies, and, of course, archeological discoveries. Even so, the farther back we look, the more scattered and ambiguous the clues become.

Anthropologists suspect that even the earliest hominids – the first members of the human family – had some sort of temporal awareness, long before our own species, *Homo sapiens*, emerged as the dominant creature on our planet.* Those early hominids, living several million years ago, "probably had a rudimentary conception of time similar to our own," argues John Shea of the State University of New York at Stony Brook. They had "an understanding of the past, an understanding of the future – and the ability to perceive the future in terms of contingencies, in terms of 'if *this*, then *that* will happen.'" Shea didn't use the word "consciousness" – the word still carries a lot of baggage in many scientific disciplines – but it seems reasonable to assume that a creature that is conscious of itself and its environment would also have at least a rudimentary awareness of time. Early hominids had enough of an awareness of past and future to live in cooperative social groups and to hunt large animals across a harsh and varied landscape, Shea says. They could learn from the past and try to predict future events; they could mentally sort through different courses of action and imagine what kinds of results they would produce. (Psychologists call this "mental time travel," and we will look at it more closely in Chapter 5.)

Evidence that those early human ancestors could plan for the future can be found around the edges of ancient lake beds in Africa and the Middle East, where archeologists have found numerous large deposits of stone tools crafted by early humans, seemingly stockpiled in strategic locations around the landscape. Perhaps, Shea suggests, they planned such caches so that the tribe would always be within a short distance of material that could be turned into weapons if the need arose. Even the sophistication of the tools suggests some degree of planning: carefully

---

* The taxonomy of early humans and related species is changing. In this book I use the term *hominid* to mean all members of the human family, including *Homo sapiens* and their extinct bipedal cousins – roughly speaking, all the primates that have ever walked upright, beginning approximately 4 million years ago. Some anthropologists now use the term *hominid* more broadly so as to include the great apes – this is, in fact, the new technical definition – and would use the term *hominan* to refer specifically to human and human-like species. For simplicity, I use the term *hominid* in its more traditional sense, which is also how it continues to be used in the popular press. (An anthropologist friend admits that the new definition causes her "no end of grief.")

chipped hand axes were almost certainly intended not for a single carving session but for repeated use. It seems that these hominids understood, in some way, past and future; they sensed that survival meant knowing not only what was over the next hill, but also what was to come the following day or the following season. As rudimentary as it may have been, they clearly had some conception of time.

One of the most intriguing – and controversial – early human behaviors is the ritual burial of the dead, a practice that emerges only in the most recent chapter of the hominid story. Such practices at least hint at our ancestors' conception of life and death, and, perhaps, of "eternity." The first signs of systematic burial can be seen about 100,000 years ago with the Neanderthals, an offshoot of the human family tree that lived in Europe and western Asia. Modern *Homo sapiens*, however, had far more elaborate burials. Before we examine such practices in detail, it's worth looking at the complex relationship between these two branches of the human family.

A Meeting of Minds

The Neanderthals make their first appearance around 130,000 years ago, and thrived until about 25,000 years ago. They lived in the same geographical area as *Homo sapiens* at the same time – they certainly overlapped in southwestern Europe during the Upper Paleolithic period, beginning around 40,000 years ago – and shared many of the same traits. Yet there would be no confusing the two species. Neanderthals were stocky and muscular, with sloping foreheads and pronounced browridges. Even if given a shave and dressed in modern clothing, a Neanderthal would draw gasps on the streets of any twenty-first-century city. They actually had bigger brains than modern humans – though size, as we shall see, is not everything.

Some Neanderthal traits seem remarkably human: they crafted stone tools, learned to control fire, enjoyed a meat-heavy diet, and cared for their elderly and sick (as revealed by the bones of individuals who lived on for many years in spite of severe disability). "There could be no more compelling indication of a shared humanity," anthropologist Richard Klein has put it. (Recently there has been a debate as to whether the Neanderthals and *Homo sapiens* occasionally interbred; the emerging consensus, based on DNA studies as well as the fossil record, seems to be that such matings happened rarely, if ever.) What is clear, however, is that their lifestyles – and their mental capabilities – were worlds apart.

The Neanderthals left no evidence of art or jewelry; although they made stone-tipped spears and could flake axes from stone, they produced very few distinct types of tools and rarely worked with bone or ivory. And there is no evidence of innovation: they made pretty much the same kind of tools for 100,000 years. Klein concludes that the Neanderthals died out "not simply because they didn't behave in a fully modern way, but because they couldn't." By contrast, early humans were prolific artists, painters, and sculptors.

The contrast between the two groups is just as sharp when we look at their burial practices. Neanderthal burials are typically just shallow pits that seem to lack any unambiguous "grave goods" or other evidence of accompanying ritual. Such burials could simply have been a hygienic way of disposing of a corpse. Only with the rise of modern humans do we see clear evidence of grave goods – tools, jewelry, and other items that would presumably be of use in the next life. Interestingly, even though *Homo sapiens* first appear on the scene in Africa around 200,000 years ago, such elaborate burial practices seem to originate only in fairly recent times, beginning perhaps 50,000 years ago – by which time our ancestors were burying their dead with great fanfare.

Life and Death – and Beyond

One of the most elaborate early human burials was found at a site called Sungir, in Russia, dating to 28,000 years ago. The bones of an elderly man rest there, alongside two adolescents, one male and one female. Each body is decorated with thousands of ivory beads (presumably attached to clothing, long since decomposed). The man wears an ivory bracelet showing traces of black paint. The boy wears a belt; under his shoulder is an ivory sculpture of a mammoth. By his right side is a huge lance, carved from a mammoth tusk. The girl wears a beaded cap; by her side are many small ivory knives or daggers.

It is hard to escape the conclusion that these citizens of the Upper Paleolithic expected *something* in the next world. As the archeologist Steven Mithen notes, they were the first beings whose world view suggests a belief in supernatural beings and, possibly, an afterlife. Their elaborate burial rituals can be seen as "the first appearance of religious ideologies," he says. Of course, *religion* is a difficult word to define, but for Mithen (and probably most scholars) it includes the religious person's assumption that death is not final. They must have believed that some nonphysical component of a person can survive after death – and that such a being could still hold beliefs and desires just like a living person. In other words, our Paleolithic ancestors had a mental picture of time that was complex enough to allow for the possibility of life after death; they could imagine time extending from this world to another, unseen world.

Within about ten thousand years of the arrival of modern *Homo sapiens* in Europe, the last of the Neanderthals disappeared. The newcomers had some kind of advantage in the struggle for survival. Many anthropologists believe that the capacity for language is what gave them – us – the edge. While other hominids, including the Neanderthals, may have grunted and gestured, modern humans developed a complex, symbolic language. With speech came the power of abstraction – a way of looking beyond the here and now. *Homo sapiens* were thinking, strategizing hunters with a sophisticated sense of time and place.

With that consciousness of time, however, came an awareness that the span of one's own life was finite. As the historian J.T. Fraser puts it, our knowledge of time "made for a double-edged weapon that cuts both ways." Our ability to plan for the future allowed our species to flourish, but, he adds, "these advantages were paid for by a profound sense of restlessness, rooted in the certainty of passing and death."

Bone of Contention

Just how carefully did early humans track the passage of time? A few calendar-like artifacts from the Paleolithic era have been put forward, but by far the most compelling is a carved bone tablet – part of an eagle's wing – discovered in a cave in the Dordogne Valley in southwestern France. The fragment – one of the most intriguing of all prehistoric artifacts – is about ten centimeters long and dates back some thirty thousand years. On its surface are a series of notches, set down in rows of fourteen or fifteen, in a winding, snakelike pattern. The American archeologist Alexander Marshack, who studied the carving in the 1960s, suggested that the notches were tally marks – that the Paleolithic hunter who made the markings was counting something. But counting what? The number of notches on each row, Marshack realized, was roughly the number of days from new moon to full moon and vice versa (the average length of a complete lunar cycle is about 29.5 days). The tablet, he speculated, may have been a primitive lunar calendar.

Anthony Aveni, of Colgate University in upstate New York, who has written extensively on timekeeping in prehistoric societies and non-Western cultures, is intrigued – and seems nearly convinced – by Marshack's claim, referring to the bone tablet as "a beguiling and captivating little artifact." Aveni acknowledges that the markings are open to other interpretations: they could have been made by a hunter keeping track of kills, or a woman tracking her menstrual cycle – or it may simply have been a knife-sharpening tool. Anthropologists also wonder if Paleolithic humans had quite enough mental agility to set up and maintain a calendar over a period of several months. Still, Aveni suspects that Marshack's interpretation is the right one: "I believe that what we have here is one of the earliest records of the passage of time, most notably the phases of the moon," he says.

While the Dordogne bone (allegedly) covers only two and a half calendar months, Aveni notes that the tally marks could in principle have been easily extended; a longer series of marks could have led early humans to understand that the period from human conception to birth was nine moons, or that certain plants and animals would become scarce at certain intervals; perhaps they could even see that the seasons would complete a full cycle every twelve or thirteen moons. Still, Aveni urges caution: "Fitting an artifact to our impression of who made it is one of the most speculative areas in the discipline of archeology."

By the late Neolithic period – around the fourth and third millennia B.C. – that interest in time's passage would be embodied in some of the most spectacular and intriguing structures on the face of the planet.* Across Europe, from the western Mediterranean to the British Isles and along the northern Atlantic coast, great stone monuments – "megaliths" – begin to appear, including dozens of stone circles, especially in Britain and Ireland. We have already had a brief look at Newgrange, one of the earliest sites believed to have calendrical significance. Stonehenge, of course, is even more famous, while nearby Avebury is larger and more complex, and Callanish, on the island of Lewis in northwestern Scotland, rivals Stonehenge in size and sophistication. These monuments have often been interpreted as observatories used to help track the motion of the sun, moon, and stars. Such interpretations invite controversy – the "'o' word" always does – but the most basic claims, such as the alignment of Stonehenge's main axis with the midsummer sunrise, are beyond dispute. (Older structures, such as the "long barrows" that pepper the countryside of southwestern England, also typically display a solar orientation – though a much less precise one. Most are oriented in an east-west direction, with the entrance pointing to the eastern quadrant of the horizon. But the actual orientation angles of these structures cover quite a wide range; some scholars have suggested a lunar rather than a solar orientation.)

---

* Cultural eras such as "Paleolithic" and "Neolithic" are not absolute; they start and end at different times in different geographical regions. The dates mentioned above refer to the late Neolithic in Europe.

The Mystery on Salisbury Plain

Stonehenge itself can be traced back nearly five thousand years, to the erection of a vast circular embankment and ditch, about 100 meters across. Just inside the embankment is a ring of fifty-six chalk-filled holes that archeologists believe once held upright wooden posts. Construction continued in spurts over the next few centuries, reaching a climax between 2400 and 2100 B.C. with the erection of a great circle of forty-tonne standing stones – also called "sarsen" stones, after the dense glacial rock used for the purpose. The sarsens are capped with ten-tonne horizontal "lintels" – an effort representing millions of work-hours. Just inside the sarsen circle is a smaller concentric ring of upright "bluestones," some of which, archeologists have determined, were

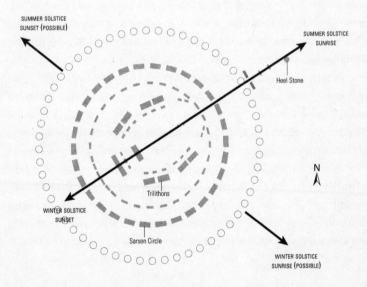

At Stonehenge, in southwestern England, one celestial orientation is unambiguous: the main axis of the central "horseshoe" grouping of trilithons is aligned with the summer solstice sunrise (or, equivalently, the winter solstice sunset). Other alignments have been suggested, but remain controversial.

dragged from the Preseli Mountains in Wales, more than 200 kilometers away.

The innermost ring at Stonehenge, consisting of five enormous trilithons (each made up of two uprights and a lintel), is laid out like a horseshoe, with its axis of symmetry lying along a southwest-northeast line. That axis also passes by an outlying stone known as the "heel stone." One can speculate that a priest or chieftain observing from the center of the monument could have used the heel stone as a kind of gun sight to observe the rising sun on the morning of the summer solstice, the day on which the sun reaches its northernmost position on the eastern horizon. (Equivalently, the observer could have been standing at the heel stone, looking southwest, watching the sun set on the evening of the winter solstice. Or, quite possibly, both.) At the very least, the monument appears to have been used as a kind of reference frame to monitor the movement of the sun and, perhaps, the moon and stars. There is no doubt that Stonehenge and the other megaliths owe *some* of their design and motivation to sky events – the question is how much, and what other motivations may have been equally important.*

In the 1960s and '70s, a few authors went to great lengths to expound on the astronomical use of Stonehenge and other Neolithic monuments – claims which were highly controversial. A few eager writers claimed that Stonehenge was a sophisticated observatory, that it served as an analog calculator that allowed – perhaps by making use of the fifty-six posts in those postholes – for the prediction of eclipses. That episode of unbridled celestial enthusiasm, as archeologist and archeo-astronomer Clive Ruggles has put it, "forms one of the most notorious examples known to archaeologists of an age recreating the past in its own image." (In the era of the Apollo moon landings, perhaps we longed

---

* On a side note, it is interesting that the site is claimed by modern-day Druids – a Celtic religious sect – as their own, considering that Stonehenge was built long before the Celtic invasion of the British Isles. Druids may have used Stonehenge, but they certainly did not build it.

to think of our ancestors as having had similar cosmic aspirations.) There is, Ruggles says, "no reason whatsoever to suppose that at any stage the site functioned as an astronomical observatory – at least in any sense that would be meaningful to a modern astronomer."

Heavenly Designs?

The problem is one of design versus chance: certain stones may be aligned with certain astronomical phenomena, but that doesn't necessarily mean that they were built with such alignments in mind. With enough stones and enough celestial "targets" – the rising or setting position of a particular bright star, for example – alignments become inevitable. "Statistically, the odds are in favour of a good celestial sightline occurring fortuitously in almost any circle," writes archeologist Aubrey Burl. Taking a site known as Grey Croft in Cumberland, England, as an example – a circular monument consisting of 12 stones – he finds there are so many possible lines and so many possible targets "that to discover nothing would be improbable." At a 12-stone circular site, he calculates, there are 132 possible alignments.

One also has to look at each site in the context of the larger landscape – both natural and artificial – examining all of the monuments in a particular region. Consider, for example, the Drombeg stone circle in County Cork, Ireland. The main axis of the circle – as at Stonehenge – aligns with the sun's solstice position. Yet there are some fifty stone circles in southwest Ireland, and none of the others seems to share such an alignment. If the builders were thinking astronomically, why was the solstice alignment the exception rather than the rule?

At Stonehenge, the solar alignment is clear enough – but the role of other celestial bodies is much less certain, and expert opinion is divided. Burl believes that extreme risings and settings of the moon, as well as of the sun, must have been recorded.* Aveni agrees that the site's main axis could have been used to monitor the most northerly moonrises during the winter months, and, in fact, he doesn't rule out its use as an eclipse indicator. When the moon rose within the stone gateway to the northeast of the circle (today, only the heel stone remains), those early skywatchers may have known that there was a possibility of an eclipse during the next full moon. "But even if no eclipse took place," Aveni writes, "that special midwinter full moon, rising opposite the setting sun, would provide ample light for a night-long ceremony to honor the attending gods." The builders of Stonehenge may well have been tracking celestial bodies, Aveni says. But, he adds, "I am convinced that if Stonehenge has anything to do with lunisolar astronomy, the association between its Neolithic builders and the sky is more closely allied with theater than with exact science." No wonder that, in the Middle Ages, Stonehenge was known as *Chorea Giganteum* – the Giant's Dance.

The Temple of Time

There can be no doubt that people once gathered at these monuments – but gathered to do what? To observe the heavens? To mark the seasons? To worship the sun and moon? To deify ancestors, or honor the dead? Very likely all of these, and more. Stonehenge was, among other things, a cemetery: recent archeological work has found that the cremated remains of more than two hundred individuals were buried within the site. Yet religious observance is not easily disentangled from cosmology – especially when celestial bodies themselves were almost certainly objects of devotion, with the sun and the moon at the top of an elaborate cosmic

---

* The moon's northernmost rising and setting positions are slightly farther north than those of the sun, while its southernmost rising and setting positions are farther south by the same amount.

hierarchy. The monument was likely home to a pantheon of gods and of spirits of animals and people. And we must remember that sites such as Stonehenge served as a meeting place for more than a thousand years; no doubt its function evolved over the centuries. We can think of Stonehenge, Aveni says, "as a place of social gathering, of religious assembly, as a cultic center, as a place of fortified habitations, a celestial temple, and observatory. All of these definitions crosscut one another, some perhaps being stressed more at one time than another."

Whatever the symbolic meaning of monuments like Stonehenge may have been, there is no doubt that those symbols involved time as much as space. Their construction suggests an interest in temporal matters far beyond what would be needed to merely track the seasons. The many burials found at Stonehenge can be seen as "a reference to the past and perhaps to mythic beginnings," according to Clive Ruggles and colleague Joshua Pollard. These were places where time may have been seen as standing still – a feeling reinforced by the continuity of celestial rhythms and public ritual. "Stonehenge always embodied notions of time – both of time past and continuity – in a world of punctuated social change," the scholars write. As Alasdair Whittle observes, the monument served as a "timeless frame of reference," a mystical arena that "made the future possible by suspending the past." It was a place where people felt united with their ancestors, the gods, the earth, and the heavens; a place where participants felt they could transcend time.

On the European mainland, equally intriguing sites are being unearthed. Archeologists have recently begun to excavate a Bronze Age settlement near the town of Goseck in eastern Germany. The site includes a circular mound and ditch, about seventy-five meters in diameter. Its origins are uncertain, but it was likely first used during the late Neolithic period, around 5000 B.C., making it significantly older than Stonehenge. Archeologists believe it was a site of cult worship but also had astronomical significance, with the "gates" to the complex aligned with the summer and winter solstices. The most arresting find: a depiction of the heavens, embossed in gold, on a bronze disk about thirty centimeters in diameter. This "map" dates from much later – about 1600 B.C. – and shows the sun, crescent moon, and thirty-two stars (possibly including the Pleiades star cluster). Archeologist Harald Meller describes it as "without a doubt the earliest genuine depiction of the cosmos." It suggests that the site where it was found "almost certainly functioned . . . as an astronomical observatory, like Stonehenge in Britain."

Fascination with celestial rhythms can be seen far beyond northwestern Europe. The ancient Egyptians and Babylonians – we'll look more closely at their achievements in the next chapter – developed a sophisticated astronomy employing mathematical and geometrical methods. In Central America, the great Mayan pyramids were aligned with the sun's position at the equinoxes, and their builders developed a spectacularly complex calendar system (more on that, too, in the next chapter). Farther to the south, the Inca were building solar observatories by A.D. 1500. In the Ohio and Mississippi valleys of North America, great earthen mounds – likely fortifications as well as ritual sites – also appear to have had astronomical significance. The Hopi, living in the southwestern American desert, used their local environment as a calendar, tracking the sun's changing position along the horizon over the course of the year. In Africa, a circle of nineteen basalt pillars in northwest Kenya may have had astronomical functions and is still used by modern inhabitants to mark important dates in their calendar. Neolithic burials in northern China, dating from 5000 to 3000 B.C., are aligned with the cardinal directions of the compass. And the list goes on.

Archeologists continue to uncover ancient stones and structures that illuminate our preoccupation with celestial cycles. In 2007, a team working in Peru's coastal desert announced the discovery of what may be the oldest astronomical site in the Americas – a series of thirteen stone structures known as the Towers of Chankillo. Dating from 1300 B.C., the towers run in a north-south line along a low ridge, with the entire complex extending three hundred meters in length. Archeologists believe the towers were used as horizon markers for observing the sun. To the east and the west of the line, they uncovered the ruins of a number of ceremonial buildings; these sites, they believe, served as observation posts. As seen from two suspected observation points, the north-south spread of the towers along the horizon matches the range of the sun's rising and setting positions over the course of the year. At the solstices, the sun would rise or set above the northernmost (or southernmost) tower; at other times of the year, the towers would have provided a way of tracking the sun's position to within an accuracy of a couple of days.

The Neolithic Mind

Nearly every claim made about prehistoric people and their conception of the cosmos is fraught with controversy. But it is clear that by the later Neolithic period – a time when anatomically modern humans had spread across the globe, hunting, farming, and tending crops – our ancestors were captivated by the night sky and its rhythmic cycles. The clockwork of the heavens had stirred something deep within us.

We will, of course, never know exactly what motivated those Stone Age builders. No doubt every archeologist has fantasized about using a time machine to drop by the site of Stonehenge during its construction. (One would feel an urge to approach the workers with a long list of questions – though it would presumably be less disruptive to watch the proceedings from behind some distant shrubs, with a pair of binoculars and a notepad.) Yet we must settle for what the stones themselves – at Stonehenge and across the Neolithic world – can tell us.

Back at Newgrange, Claire Tuffy has often entertained those kinds of thoughts; she can't help wondering what was going through the minds of the monument's builders as they labored over the massive stones and carved those mysterious figures. Our best guesses, she says, will almost certainly remain just that. She also recognizes that we can't help reading into the stones any number of motivations and desires that may or may not have been shared by the monument's builders. In the twenty years that she's been working at the site, Tuffy has heard visitors describe their experiences in terms that often reflect, more than anything, our ever-changing pop-cultural visions of the universe. In the 1970s, in the wake of Erich von Däniken's *Chariots of the Gods*, visitors would sometimes tell her that the mound looks like a spaceship. (The idea that aliens from another world built Newgrange used to annoy her – "They were going to give the credit to some guys from outer space," she laments.) In the 1980s, as the environmental movement gained ground, the earth itself became a revered entity; people started dowsing for "earth energies" and spoke of living in harmony with "Mother Earth." These days, Tuffy says, many visitors are thinking in terms of spirituality and a yearning for a universal religion. "In a world where we're abandoning established religion, people are branching off on their own, and they're going back to these places to find answers."

We take a last look at the spiral patterns on the rocks in the back of the tomb. Archeologists suggest that the markings symbolize the sun – a plausible enough idea, and one that seems to mesh with the site's winter-solstice solar orientation. But what, Tuffy asks, did such symbolism really mean to the builders of Newgrange? "We don't bring the same meaning with us as they would have," she says. "What did the sun mean to the people five thousand years ago? That's the big gap that will never be bridged."

Exiting the cave, we both squint in the morning sunlight.

"Even if we got the old time machine and zipped back to 3100 B.C.," Tuffy speculates, "I'm not quite sure that we'd be able to communicate adequately to understand them, to see the world like they see it."

YEARS, MONTHS, DAYS

The quest for the perfect calendar

To everything there is a season, and a time to every purpose under the heaven.

– Ecclesiastes 3:1

Every fall – or nearly every fall – talk-show host David Letterman dips into his bag of tried-and-true seasonal jokes and pulls out a nugget that goes something like this: "So, Happy New Year to all our Jewish friends out there. Today is Rosh Hashanah, the Jewish New Year. This marks the beginning of the year 5769 in the Jewish calendar" – as it would be in the 2009 version of the joke. "Now, are you like me? Are you still writing 5768 on your checks?" (Cue rim-shot from the drummer, and a guffaw from bandleader Paul Shaffer.) Of course, no one – not even Orthodox Jews living in Jerusalem – writes 5768 or 5769 on their checks. Instead, everyone from Seattle to Singapore uses – for civil purposes, at least – the Gregorian calendar, a remarkable invention that incorporates the ideas of the Babylonians and the Egyptians, modified by the Romans, and polished into its modern form by a sixteenth-century pope and a team of nearly forgotten astronomers and mathematicians. The Gregorian calendar is one of the most successful ideas in the history of civilization. (Richard Dawkins might call it a successful "meme" – a unit of cultural information that propagates over time.)

The Gregorian calendar is not the only timekeeping system invented by humankind – nor, as we'll see, is it even (by some measures) the most accurate. But its story is a noteworthy one, an achievement centuries, even millennia, in the making. We saw in the previous chapter how early humans were captivated by – and began to follow – the regular motions of the night sky. By the time of the great ancient civilizations, such systematic observation had become a virtual industry; every culture would develop some sort of calendar for mapping out the year, based on their observations of the heavens and their own particular needs and priorities. The one that rules today – the Gregorian Christian calendar – exploits ideas from many different cultures, each with a unique perspective on the significance of the heavenly bodies and unique solutions to the problem of tracking their motions. In this chapter we'll take a look at some of the challenges confronting would-be calendar makers through the ages, as they tried to tame the myriad of motions displayed by the sun, moon, and stars.

The first rudimentary steps toward tracking those celestial motions, as we've seen, may have occurred as early as the Paleolithic period. But it is only with the rise of the first civilizations – marked by complex, agriculture-based urban settlements with full-blown writing systems – that we can be certain that people were keeping track of days, months, and years. Making sense of those celestial cycles, however, is complicated by the fact that neither the number of days in the lunar cycle nor the number of lunar cycles in a year is a nice round number (indeed, not even a whole number). The lunar month, as mentioned earlier, is about 29½ days long (actually 29.5306); the average solar year (also known as the "tropical" year) is about 365¼ days long (actually a smidgeon less, at 365.2422 days). That these cycles did not fit neatly into one another was well known: back in the fifth century B.C., the Greek poet Aristophanes, in his play *The Clouds*, had the moon complaining that the days refused to keep pace with her phases.

Incongruent Cycles

Try dividing the length of the year by the length of the lunar month, and again you get a fractional number, greater than 12 but less than 13 – the true figure is close to 12.3683. Over the millennia, different civilizations tried every possible trick for reconciling these incongruent cycles. Some simply rounded the length of the month up to 30 days, a practice adopted by the ancient Sumerians; 12 such months yield a 360-day year, just 5 days (roughly) short of the true solar year. Others used a more precise length for the lunar cycle and then assumed there were exactly 12 months in a year: the result is a year that is 354 days long – 11 days short (roughly) of the true solar year. Adopt such a calendar, and each New Year's celebration will be 11 days earlier than it was the year before. A midsummer celebration would become a midwinter celebration after just 16 years.

Any calendar system that uses the phases of the moon to track the months but also attempts to reconcile those months with the cycle of the seasons is called a *luni-solar* calendar. The Babylonians adopted one such system. A new month was determined by the first sighting of the crescent moon in the western sky – a practice that continues in Muslim nations to this day (notice how many Muslim nations feature the crescent moon on their flag). To keep the months in step with the solar year, the Babylonians employed a cycle in which seven 13-month years alternated with 12 years of just 12 months. The result was a 19-year cycle known as the Metonic cycle, after the Greek astronomer Meton of Athens, who lived in the fifth century B.C. (Meton discovered that 235 lunar months amount to almost exactly the same interval as 19 solar years; a calendar based on this cycle would deviate from the true solar year by just 1 day every 219 years.)*

---

* As is so often the case, the credit did not go the person who first came up with the idea; besides the Babylonians, the Chinese recognized the nineteen-year cycle centuries before the time of Meton. In just the same way, the "Pythagorean theorem" was almost certainly known to the Babylonians a thousand years before the Greek thinker gave his name to it.

Beginning in the second millennium B.C., the extra month would be added – "intercalated" – following either the sixth month (Ululu) or the twelfth month (Addaru) of the Babylonian calendar. We have a record dating from the nineteenth century B.C. of King Hammurabi's decree on just such an adjustment:

> This year has an additional month. The coming month should be designated as the second month Ululu, and wherever the annual tax has been ordered to be brought in to Babylon on the 24th of the month of Tashritu it should now be brought to Babylon on the 24th of the second month of Ululu.

The Jewish calendar is closely modeled on the Babylonian. (The mutual influence of the two cultures can be traced back to the sixth century B.C., when Babylon, under Nebuchadnezzar II, conquered Jerusalem; the Jewish people spent the next 70 years or so in exile.) The Jewish calendar, like the Babylonian, is built on the nineteen-year Metonic cycle, with its combination of 12-month and 13-month years. Within that cycle, the lengths of certain months can also vary, so that a "regular" year can be 353, 354, or 355 days long, while a leap year (containing an extra month) can be 383, 384, or 385 days long. (This is why the date of Jewish holidays such as Hanukkah leaps around so much with respect to the Gregorian calendar.)

The Rhythm of the River

The ancient Egyptians had different priorities from the Babylonians: in the Egyptian year, the most important event is the annual late-summer flooding of the Nile River. The floods bring life to the desert; the river is the focal point of Egyptian civilization. It is no wonder that the Greek historian Herodotus, describing the geography of Egypt in the fifth century B.C., referred to "the gift of the Nile." The Egyptians could predict the annual flooding by carefully tracking the brightest star in the sky, Sirius (which they equated with the god Sothis, a serpent). Every spring, Sirius disappears for several weeks, hidden by the sun's glare; to the Egyptians, this was the serpent's journey through the underworld. The New Year was at hand when Sirius first became visible again in the pre-dawn sky (in modern astronomical jargon, its "heliacal rising"); its return heralded the imminent flooding of the Nile. The importance of Sirius in Egyptian astronomy is reflected in its grandest monuments, including the Great Pyramid of Cheops in Giza. The narrow shafts that lead into the central chambers are believed to be aligned with the path that Sirius traces across the sky. (We've retained at least one colloquial saying from this bit of calendrical lore: the return of Sirius, the "Dog Star," heralds the "dog days" of summer.)

With the annual flooding as their first concern, the Egyptians aban-doned the moon in favor of a solar calendar. They still used months – 12 of them, each 30 days long – but their months rolled along independ-ently of the actual phase of the moon. That, of course, adds up to only 360 days – so they added 5 days for religious celebration at the end of the year. The resulting year of 365 days is only about a quarter-day short of the true solar year. Interestingly, the Egyptians recognized that slight shortfall very early on; they realized that adding a quarter-day to the year (or an additional day every four years) would bring the calendar within just a few minutes of the true solar year. Yet centuries would pass before the change was adopted by the priests in charge of the calendar system. It wasn't until 238 B.C. that the Egyptian king Ptolemy III urged the adop-tion of a leap-year system. (Egypt's rulers at this time were Hellenistic Greeks, and Ptolemy himself – not to be confused with the famous astronomer who lived many centuries later – may have been advised on astronomical matters by Aristarchus of Samos, the noted Greek astronomer and mathematician who promoted the idea of a heliocentric universe some 18 centuries before Copernicus.) However, another two centuries would pass before the leap year was adopted. The conquest of Egypt by Rome finally forced the issue; the emperor Augustus made the Egyptians adopt the leap year to bring the Egyptian calendar in line with the Julian calendar used in Rome.

Caesar's Calendar

The Romans, like the early Egyptians, initially used a calendar of twelve lunar months, with extra days or months inserted from time to time to keep in sync with the seasons. The system was far from perfect: as writer David Ewing Duncan explains, it suffered both from neglect and from political manipulation. The priests in charge of the calendar, he writes, "sometimes increased the length of the year to keep consuls and senators they favoured in office longer, or decreased the year to shorten rivals' terms." They also used the calendar "to increase or decrease taxes and rents, sometimes for their personal financial advantage." By the time of Julius Caesar, the Roman calendar was in desperate need of reform. As the Greek historian Plutarch tells us, the Romans wanted to establish

> a certain rule to make the revolutions of their months fall in with the course of the year, so that their festivals and solemn days for sacrifice were removed by little and little, till at last they came to be kept at seasons quite the contrary to what was at first intended, but even at this time the people had no way of computing the solar year; only the priests could say the time, and they, at their pleasure, without giving any notice, slipped in the intercalary month.

Caesar's reform was "projected with great scientific ingenuity," Plutarch writes, for the emperor "called in the best philosophers and mathematicians of his time to settle the point," in the end adopting "a new and more exact method of correcting the calendar." Under the new system, Rome succeeded "better than any nation in avoiding the errors occasioned by the inequality of the cycles." At the heart of the Julian reform was the same idea suggested by Ptolemy III two centuries earlier – the adoption of a leap year every four years. Three out of every four years would be 365 days in length; the fourth would be 366 days long (giving an average of 365¼ days). The Julian year consisted of alternating 30-day and 31-day months – months that started and ended independently of the phase of the moon – with only February as the oddball (with 28 days in a regular year and 29 in a leap year). To correct for the "drift" that had already accumulated up to that time, Caesar ordered two extra months inserted into the year that we would label as 46 B.C. – making that particular year some 445 days long. He called it the *"ultimus annus confusionis"*: "the last year of confusion."

The new, improved calendar was more than a scholarly pursuit for Rome's elites. It also "injected a new spirit into how people thought about time," according to Duncan. Until the Julian reform, time

> had been thought of as a cycle of recurring natural events, or as an instrument of power. But no more. Now the calendar was available to everyone as a practical, objective tool to organize shipping schedules, grow crops, worship gods, plan marriages, and send letters to friends . . . The new Julian calendar introduced the concept of human beings ordering their own individual lives along a linear progression operating independent of the moon, the seasons, and the gods.

It is interesting to compare this "linear progression" to other conceptions of time from other cultures. What is emerging in Rome at the time of Caesar seems to be a uniquely Western sense of time, an image of time as something like tick-marks on a meter stick – very much like the mental picture most of us have today when we look at our watch or jot down an appointment in our calendar. This is a profound notion, and we will discuss it more in the chapters ahead.

It was at around this time that the people of Rome began to mark the New Year in January rather than March, probably in an effort to bring the date closer to the winter solstice. The Senate would eventually change the name of the seventh month (the fifth – Quintilius – in the old system) to Julius (July) in Caesar's honor. The emperor Augustus would later fiddle with the length of the months a bit more: if July was to have thirty-one days, so, surely, must August. The result did not change the length of the year, but it did give us the somewhat random distribution of month-lengths that we have today.

Centuries later, after the Roman Empire had adopted Christianity as its official religion, a monk named Dionysius Exiguus ("Dennis the Little," *c.* A.D. 470–544) gave the calendar a new starting point, labeling the years after the birth of Christ as A.D. – *anno Domini,* "the year of our lord." Of course, Dionysius could only estimate the year of Christ's birth; modern historians peg the event to 4 or 5 B.C. It also did not occur to Dionysius to include a year zero – the concept of zero had not yet taken hold in the West – and so the year 1 B.C. would be followed immediately by A.D. 1. (Incidentally, the term B.C., for "before Christ," was adopted only in the early seventeenth century. It is interesting to note that, while we abbreviate a Latin phrase for "A.D.," we abbreviate an English one for "B.C."; by the time the latter notation came into regular use, English was already beginning to replace Latin as a language regularly used by educated people.)

We've seen how the Romans eventually moved the New Year to January 1, a move that would gradually be adopted across the Western world (although it did not happen quickly – Britain and its colonies adopted January 1 only in 1752). But the choice of when to celebrate the New Year is, after all, an arbitrary decision. Many cultures took the spring, with its suggestion of rebirth and renewal, as the starting point. In South America, many native cultures used a heliacal rising, as the Egyptians did – but with the Pleiades rather than Sirius as the focus of attention. (In many native languages, the word for "year" and "Pleiades" is the same.) The key role played by the Pleiades can be seen in some of the surviving ancient structures of the region, especially those of the Inca. At Machu Picchu, for example, an oval-shaped stone building known as the Torreon has one of its windows aligned with the point on the horizon where the Pleiades rise. (The recently discovered Incan city at Llactapata, a "suburb" of Machu Picchu just across the Urubamba River, features buildings with the same orientation; its temple and observatory are aligned with the sun's position at the solstices and the equinoxes, as well as with the Pleiades.)

Mind you, the calendrical signals that determine the start of the New Year need not be celestial. For the Trobriand Islanders of the western Pacific, the year begins on Worm Day, when the palolo worm begins to spawn – generally between mid-October and mid-November.

Keeping Time by the Moon

The lunar month – the period from one new (or full) moon to the next – is perhaps even more obvious than the yearly cycle of the seasons; we saw in the last chapter how it may have inspired calendar-like tallying as early as the Paleolithic era. At the very least, the start (or end) of the lunar cycle can be observed with relative ease. Some cultures, such as those that embrace Islam, have based their calendar on the moon exclusively. The Islamic quest for accurate timekeeping – a requirement for their strict regimen of prayer – led Muslim nations in medieval times to become experts at astronomy. By the turn of the first millennium A.D., Muslim scientists had perfected astronomical instruments like the astrolabe and established great observatories across the Middle East. The oldest surviving Muslim observatory is the recently renovated structure at Maragha (present-day Maragheh) in northern Iran, dating from the thirteenth century. The Muslim year is strictly lunar, consisting of 12 cycles of the moon; the length of a "month" is alternately 29 or 30 days. The "year" runs only 354 days in length – some 11 days short of a solar year. For that reason, the Muslim year, including all the Muslim holidays, drifts relative to the seasons. Arab scholars were, nonetheless, well aware of the true length of the solar year. The scholar and poet Omar Khayyám (c. 1048–1131) calculated the length of the solar year at 365.24219858156 days; this is an accurate figure – the year is about 365.2422 days long – but, as Duncan puts it, Khayyám's calculation is "overly precise" because of the irregularity of the earth's rotation (meaning that all those extra decimal places have no practical value). Khayyám also worked out a calendar with 8 leap years every 33 years – a somewhat awkward system, but still more accurate than our Gregorian calendar.

Any discussion of calendars and ancient civilizations must make special note of the accomplishments of the Maya of Central America. In their quest for – one might say obsession with – accurate timekeeping, the Maya piled cycle on top of cycle: they marked not only the (approximately) 365-day cycle of the seasons, but also a slightly shorter cycle of

360 days known as the *tun* and a 584-day cycle associated with the movement of the planet Venus – for them, the planet associated with warfare. Perhaps even more fundamental to their way of tracking time was a 260-day "sacred round." The full rationale behind the Mayan fondness for this number will probably never be known, but Anthony Aveni suspects it is linked to its compatibility with so many other cycles in nature. As Aveni points out, 260 is roughly equal to the average number of days that Venus spends in the evening or morning sky (actually 263 days), and is very close to the average human gestation period (253 days). It is also, roughly, the average length of the agricultural season in many parts of Mexico. For all of these reasons, says Aveni, "the 260-day period emerged in the world of the Maya as the cycle par excellence to encapsulate the powers of all the gods – the gods of time, the sun, earth, moon, those of fertility, and rain." The number 260 could be thought of as "the Maya divine temporal common denominator."

The Mayan calendar implied a length for the solar year that is closer to the mark than ours, at 365.2420 days, just shy of the actual value of 365.2422 (the Gregorian value is 365.2425). As with Omar Khayyám's calculation for the length of the year, it is tempting to label the Mayan calendar as more "accurate" than ours, but there are some subtle reasons why we should resist making such a claim. For one thing, the earth's rate of rotation is not stable over thousands of years – more on that in the next chapter. As well, there is more to developing a calendar than estimating the length of the solar year. The scholars who developed the Gregorian calendar were also interested in keeping the date of the vernal (spring) equinox fixed from year to year, and our calendar, says astronomer and author Duncan Steel, does a better job of this than the Mayan calendar.* But such comparisons, he admits, are cases of "comparing apples with oranges."

For the Maya, the calendar and its cycles offered a mental leap across the abyss of time, a glimpse of eternity. To peer arbitrarily far into

---

* The vernal equinox marks the beginning of spring for those of us in the northern hemisphere; in the southern hemisphere it marks the start of autumn.

the past or future, you begin with the *tun* and simply expand it with multiples of 20: 20 *tuns* make a *katun* (7,200 days, or nearly 20 years); 20 *katuns* make a *baktun* (144,000 days, or more than 394 years); by the time we get to the *alatun*, we're talking about a period of some 63,000 years.

The Maya and "Organic Time"

What most distinguishes the Maya conception of time from the Western view, however, is not the variety of counting systems but the imagined nature of time itself. For us, time is *inanimate*: we feel that it passes at a constant rate, with no heed paid to man or machine. We can neither give it a boost nor slow it down. For the Maya, however, time was organic – and men and women were intimately involved in its passage. Mayan rulers, charged with keeping that temporal symphony in tune, were seen to embody the very essence of time. As David Stuart of Harvard University writes, that role "was fundamental to the cosmological underpinnings of divine kingship." In the Mayan world, "the king came to be explicitly identified with the temporal mechanisms of the cosmos."

The Maya, writes Anthony Aveni, had an "enduring obsession" with timing of events both human and celestial – and, indeed, could not conceive of a separation between the two:

> The Maya were fatalists at heart. They strove to find certain repeatable patterns recovered from the observations and recorded data of past sky events, which could then be used as a guide for predicting the future. For them, such patterns constituted realizable proof of the long-held Mayan belief that the future was present in the past – that the unfolding of events over time's near and distant horizon actually could be foretold by looking with introspection over one's historical shoulder.

The Mayan penchant for harnessing endless celestial cycles can be seen in one of their most famous surviving artifacts, the Dresden Codex.

(Written in the twelfth century, the stone tablet is named for the German city where it was rediscovered some seven centuries later.) The Codex contains a record of 205 lunar cycles, over 11,968 days. With a mathematical notation based on the number 20 (rather than our base-10 system), the Codex could be used to predict both solar and lunar eclipses – knowledge that no doubt made the ruling elite seem even more powerful.

Mayan fatalism can be seen even in their names for the days. Each of the twenty named days in the Mayan calendar was essentially deified, and said to have a distinct personality; certain days were deemed good for some tasks and bad for others. A child born on a particular day would be expected to have certain character traits associated with the day and the god in question. For the Maya, the days "took on a life of their own," says Stuart, who also curates the Mayan collection at Harvard's Peabody Museum.

Because time was organic, it was also responsive to the actions of man. In fact, keeping time on its course was a community effort; everyone had to pitch in. The king, however, had the greatest responsibility. As divinely ordained ruler, he was seen as the embodiment of time, and it was his duty to use time to maintain the social, political, and cosmological order.

The burden of that temporal responsibility comes to life in one of the Peabody's most striking exhibits – a cast of a stone monument from the Mayan city of Copán, in Honduras, known as Altar Q, which Professor Stuart showed me on a recent visit to Harvard. The square stele is carved with the figures of sixteen kings – four to a side – spanning nearly four hundred years of history. Time wraps around this monument so that the sixteenth monarch is face to face with the first. The old king is passing what looks like a torch to the new king.

"This is the guy who actually dedicated the stone," Stuart explains, pointing to the sixteenth and final king. "He's the living king. His name was Yax Pasaj." Beside him are glyphs indicating the date of his inauguration and commemorating the symbolic passing of the throne from the founding king. Stuart grins. "Isn't that cool? Time has just come full circle."

Yax Pasaj was indeed the last king; the regime soon collapsed. Perhaps it was the result of the droughts, famine, and warfare that were swirling around Copán at this time (the eighth century A.D.). But Stuart can't help wondering if the calendar played a role: the king began his reign just as the *baktun*, the 394-year cycle, was drawing to a close. Perhaps the story of Copán ends with this particular ruler "because history did come full circle," Stuart suggests. "It would have been hard for any king who came after this to really put himself into this cosmological scheme." The Maya of Copán, he says, "may have seen it as a time when change was necessary."

The Oddball Cycle: The Week

We have spoken of days, months, and years, each of which are clearly delineated by the motion of the heavens. The week, in contrast, seems to be much more artificial. The modern week is both more rigid and yet less predictable than the month or the year. While the modern (Gregorian) year can have either 365 or 366 days, and a month can have anywhere from 28 to 31 days, the fixed 7-day week seems inert. And yet, while each January 1 marks the beginning of a new year and a new month (and sometimes a new decade, century, or millennium), six times out of seven it falls haphazardly in mid-week.

The roots of the week are much less obvious than the other calendrical units. It may have originated in an attempt to divide the month into four roughly equal parts: new moon / first quarter / full moon / third quarter – although technically this would give a week slightly greater than seven days in length ($29.53 / 4 = 7.38$ days). There was another celestial motivation in favor of the number seven: in ancient times, the number of "wandering bodies" known in the heavens included the sun, moon, and five planets (Mercury, Venus, Mars, Jupiter, and Saturn) – seven bodies in all. The seven-day week may owe its origins to the Babylonians, who made an astrological connection between the gods and the days of the week.

The idea of the week also hinges crucially on the notion of the Sabbath – a special day of rest associated most clearly with the creation story in Genesis, the first book of the Hebrew Bible. The word *Sabbath* appears to come from the Babylonian *sabattu*, an evil day associated with the moon god, Ishtar. Yet the Babylonians did seem to give special status to every seventh day, and a separate Babylonian word, *sibitu*, means simply "the seventh." Most scholars agree that the seven-day week, with a weekly day of rest, is a Jewish idea, adapted with modification from the Babylonians in the years during and after the Exile.

Most other religions made connections, as the Babylonians did, between various gods and the days of the week. The modern English names for the days of the week can be traced back to Saxon names for those gods. Their Latin equivalents make it clear which god is being honored: the Latin *Dies Solis* is the Saxon Sun's day, which has become our Sunday; *Dies Lunae* is Moon's day or Monday; *Dies Martis* is the Saxon Tiw's day (Tuesday); *Dies Mercurii* is Woden's Day (Wednesday); *Dies Jovis* is Thor's Day (Thursday); *Dies Veneris* is Frigg's day (Friday); *Dies Saturni* is Seterne's day (Saturday). One cannot help wondering how often Woden, the German (and Viking) god of poetry, would come up in every-day conversation had he not been immortalized in the "hump-day" of the Western calendar.

But why do the days run in that particular order? It is not the order of brightness: if it were, Jupiter and Saturn would come before Mars and Mercury. Nor is it based on the rate at which the bodies move against the background stars over time – what astronomers call the "sidereal period." (That sequence, from longest to shortest, is Saturn, Jupiter, Mars, Sun, Venus, Mercury, Moon.) Nor is it based on the rate at which the bodies line up with the sun in the sky – the "synodic period." (That order would be Mars, Venus, Jupiter, Saturn, Mercury, Moon; the sun's synodic period is undefined.) The answer, most historians agree, involves a link between the days of the week and the hours of the day. When ancient astronomers divided the day into twenty-four hours, they associated a particular heavenly body with each hour, believing it "ruled" that hour. Saturn, the slowest-moving body, was believed to be the most powerful, and was thought to control the first hour of the first day (which was, in ancient times, Saturday). Then, using the sidereal sequence, Jupiter becomes associated with the second hour of the first day, Mars with the third, and so on. After linking the moon with the seventh hour, we cycle back to Saturn, which controls the eighth, fifteenth, and twenty-second hours (every seventh hour). Mars controls the twenty-fourth hour of the first day, and so the sun becomes linked to the first hour of the second day – Sunday. Continuing in this manner, cycling through the seven bodies and the twenty-four hours of each day, we get the modern sequence of the days of the week. Eventually – possibly in Roman times – the week gained independence from the month and the year, and Sunday replaced Saturday as the first day of the week.

It wasn't only the heavens that dictated the structure of the week. Seven days is a convenient interval for a market cycle, a chance for farmers and merchants to come together and exchange goods. The choice of seven days wasn't universal: some African tribes kept a five-day market cycle; for the Inca of South America it was eight days; it was ten for the ancient Chinese. Postrevolutionary France tried to establish a ten-day week in 1792, without success. The Romans had an eight-day market cycle in place when, in A.D. 321, the emperor Constantine established Sunday as the first day of an official seven-day week. (The Beatles song "Eight Days a Week," the author Duncan Steel quips, may represent a yearning for a Roman girl from the time before Constantine.)

Julius Caesar's calendar was a momentous achievement – but it was also deeply flawed. Its average year of 365.25 days was just slightly shorter – by about 11 minutes – than the true solar year. By the time lawyer and statesman Ugo Buoncompagni was elected Pope Gregory XIII (1502–85), that discrepancy added up to 10 full days. The Julian year was drifting with respect to the seasons, dragging all of its feasts and holy days with it. If nothing were done, Easter would eventually become a summer holiday.

The Problem of Easter

Before we examine Gregory's solution, it's worth taking a closer look at the timing of Easter, the most important festival in the Christian year. The holiday celebrates the resurrection of Jesus, which Christians believe occurred three days after his death on the cross. (Historians now place the death of Jesus sometime between A.D. 27 and 33.) The holiday is intimately linked to the Jewish Passover: the "Last Supper" shared by Jesus and his disciples before his execution is believed to have been a Passover seder, based on the outline of events described in the New Testament of the Bible. It wasn't until the second century A.D. that Easter began to be celebrated; no doubt it evolved from older pagan holidays associated with the coming of spring. (The word *Easter* comes from the Norse god of the spring, Eostre.) Different sects, however, marked the holiday on different dates. Some groups followed the Jewish calendar, celebrating Easter on the same day that the Jews celebrate Passover – the fourteenth of Nisan – regardless of what day it fell on. Other Christian groups gave preference to the day of the week, marking the holiday on the Sunday that followed.

Eventually the latter approach prevailed: Christians decided to mark the crucifixion on Good Friday and celebrate the resurrection on the following Sunday. But which Sunday? The simplest solution would have been to choose the first Sunday after Passover – but there were several perceived reasons not to do so. One involves the nature of the Jewish calendar itself, and the wide range of dates on which Passover can fall. Passover is often said to fall on the "first full moon of spring," but that is an oversimplification. Because the Jewish calendar employs whole leap months rather than merely an extra day (as in the leap years of the Julian calendar), Passover can fall on a wide range of dates compared to the Julian (and, later, Gregorian) calendars. In the longer years – those that have thirteen months – Passover can fall as much as a full month after the equinox (as happened in 2008). The Church wanted to ensure that Easter followed the equinox by no more than a month.

There were also other factors in play: Christians wanted to distance themselves from the Jews, and didn't want their most sacred holiday too closely linked to a Jewish festival. As Duncan Steel writes, the Christians "invented a reason to disagree with the Jewish system." In the middle centuries of the first millennium A.D., Church leaders worked out several methods for computing the date of Easter, all of them rigged so that Easter and Passover would never coincide.

The Easter controversy was one of the hot topics for discussion when the leading figures of the Christian world gathered in Nicaea, in what is now northeastern Turkey, in A.D. 325. (Constantine presided over the meeting himself; though he would wait until he was close to death to be baptized, he had long been sympathetic to the growing religion.) More than 300 clerics and scholars took part in the council. Exactly what they concluded with regard to Easter is somewhat obscure – the original records have not survived – but there was evidently a push to have all of Christendom celebrate Easter on the same day. Even so, the controversy lingered; some groups relied on Egyptian scholars for advice on selecting the appropriate date, while others continued to rely on the Jewish calendar.

Ultimately, the Christian authorities decided to sidestep astronomical methods, relying instead on a mathematical model that would simulate the motion of the real sun, moon, and stars. Once an accurate model was established, the date of Easter could simply be calculated; there would be no need to consult either the astronomers at Alexandria or the priests in Jerusalem. At some point – certainly after Nicaea – it was agreed that Easter would be celebrated on the first Sunday after the fourteenth day of the "paschal moon" – with the paschal moon defined as the first lunar month whose fourteenth day falls after the vernal equinox (remembering that the lunar month begins with the new moon). Clear as mud, right? For what it's worth, this is roughly – though not precisely – the same as the first Sunday after the first full moon after the vernal equinox (with the additional caveat that if that full moon fell *on* a Sunday, Easter was to be delayed until the following Sunday). These rather convoluted rules achieved at least one goal: they ensured that Easter was never celebrated on the same day as Passover. Even so, different churches still managed to disagree as to the date of Easter – in part because the bishops in Alexandria and Rome could not agree on the date of the equinox on which the whole computation depended.

The struggle over Easter, incidentally, illuminates one of the greatest misconceptions about the relationship between the Catholic Church and science. In the wake of the Galileo affair, the Church was often seen as hostile to scientific enquiry. Yet throughout the Middle Ages – and well into modern times – the Church was actually one of the strongest supporters of precision astronomy and timekeeping, as a direct result of the Easter dilemma. In fact, dozens of churches and cathedrals, including those in Rome, Milan, Florence, and Bologna, also served as observatories; many were equipped with strategically placed apertures in walls or ceilings that allowed a beam of sunlight to strike a north-south "meridian line" on the floor. Such measurements helped establish the dates of the solstices and equinoxes, on which the Easter calculations depended.

The next critical step was taken by the monk Dionysius Exiguus. Sometime in the sixth century, he worked out a set of tables that would allow the date of Easter to be computed decades, even centuries, in advance. Dionysius's tables were indeed used for many centuries – but they were inherently flawed. First of all, his value for the length of the lunar cycle was ever-so-slightly off. But secondly, and more importantly, his value for the length of the year itself – inherited from the Julian calendar – was too short, and by Gregory's time the accumulated error added up to more than a week.

Gregory's Solution

In the mid-1570s, Pope Gregory XIII convened a calendar commission to tackle the problem. The commission included a physician named Aloysius Lilius (c. 1510–76), a Jesuit astronomer named Christopher Clavius (1538–1612) – known to historians as the "Euclid of the sixteenth century" for his mathematical insights – and a handful of lesser-known players. The commission members grappled with their charts and tables, trying to deduce the true length of the year, and struggled to make some combination of regular years and leap years average out to that length. It was the doctor, Lilius, who nailed down the winning formula. He realized that the Julian calendar was slipping by about 1 day every 134 years – or 3 days every 402 years. For simplicity's sake, he rounded that to 400. Suppose, he suggested, that the new calendar drops 3 days every 400 years. In the Julian calendar, "century years" like 1500, 1600, and 1700 would be leap years because they're divisible by 4. In the new plan, only those century years divisible by 400 (such as 1600) would be leap years; other century years – which would have been leap years under the Julian system – would have just 365 days in the new scheme. (Notice, by the way, that the first year to be affected by the new proposal would be 1700 – by which time all those involved in the reform would be safely in a timeless realm.)

Lilius is lucky that the plan worked as well as it did. He based his calculations on numbers set down in 1252, known as the "Alfonsine tables." Named for the Spanish king Alfonso the Tenth, the tables presume a solar

Pope Gregory XIII presides over the Commission for Calendar Reform, *c.* 1582. *(Scala / Art Resource, NY)*

year of 365 days, 5 hours, 49 minutes, and 16 seconds – about 30 seconds longer than the true year. Lilius's proposed reform, however, gives an average year that's slightly closer to the mark, at 365 days, 5 hours, 48 minutes, and 20 seconds – just 26 seconds short of the true year. The Gregorian calendar still runs ever so slightly "fast" compared to the seasons, gaining one day every 3,300 years.

The commission also grappled with the date of Easter, finally putting the centuries-old dilemma to rest – though their solution for the holiday's computation seems, to the untrained eye, more complicated than ever. It still involves a mathematical model that simulates the moon's motion, based on the nineteen-year Metonic cycle, as well as esoteric constructs known as "golden numbers" and "epacts," which, happily, are beyond the scope of this discussion. In spite of all the arcane mathematical contortions, the date of Easter is still very nearly "the first Sunday after the first full moon of spring." The Christian calendar remains a luni-solar calendar – one that keeps pace with the seasons but that also celebrates certain holidays, such as Easter, based on the phase of the moon.

Pope Gregory was swayed by the recommendation of Lilius and the commission, and issued a twenty-page compendium outlining the reform on January 5, 1578. It included, among other things, the declaration that the New Year would be marked on January 1, as it had been for Caesar, fifteen centuries earlier. The reform was finally enshrined in a papal bull issued on February 24, 1582.

The reform also ordered ten days to be dropped from the calendar in a one-off move to make up for the days that had been lost by the use of the Julian calendar for so many centuries. And so October 4, 1582, was followed by October 15. Some people were distressed at what seemed to be "lost" time. Merchants fretted over how to calculate profits and losses; bankers were befuddled by interest rates.

Most Catholic countries came on board right away. Italy, Spain, and Portugal adopted the Gregorian reform immediately; France and Belgium joined a few months later; Catholic portions of Germany and Switzerland made the switch within a couple of years. Protestant countries, including the Protestant states within Germany, opposed the reform; as David Ewing Duncan writes, one bitter theologian called Pope Gregory the "Roman Antichrist" and dismissed his calendar as "a Trojan horse designed to trick real Christians into worshipping on the incorrect holy days." As the decades passed and more nations and peoples adopted the reform, resistance became more and more futile. By 1700, most of Germany and Denmark had complied. Sweden joined, after much hand-wringing, in 1753.

"Give Us Back Our Eleven Days!"

England was especially problematic, with a Protestant Queen (Elizabeth I) constantly embroiled with Catholic agitators – although her most trusted adviser, John Dee (1527–1608), urged the adoption of a slightly altered version of the reform. Ultimately, another 170 years would pass before England (and by then, Britain) adopted the Gregorian calendar. A bill was drafted by a retired politician and former secretary of state, Philip Dormer Stanhope (1694–1773), the Earl of Chesterfield, and read before Parliament. The bill passed, and was signed by King George II on May 22, 1751 (though, as Duncan points out, Stanhope admitted that he himself "could not understand" the details of the argument it rested on).

To catch up to the Gregorian calendar, Britain and its colonies would have to drop eleven days – the ten that the Catholic countries had added under the Gregorian reform, and one more because those countries had marked 1700 as a leap year, while Britain, still on the Julian calendar, had not. And so September 2, 1752, was followed by September 14. From then on, as well, the New Year would begin on January 1 rather than March 25, as it had until that time. Again, ordinary people balked at the "lost" days; there were riots in London and Bristol, where protesters shouted, "Give us back our eleven days!"

The Eastern Orthodox Church also rejected the reform. To this day its members celebrate Easter on a different day from other Christians around the world. However, many Orthodox countries did accept the reform for civil purposes in the early years of the twentieth century. Russia came on board after the revolution, in 1918; China held out until the Communist takeover in 1949.

It's not clear if human civilization will last another thousand years (as Yogi Berra supposedly quipped, "It's tough to make predictions, especially about the future"). But if people are still around to relish in late-night talk shows, then I suspect a thirty-first-century David Letterman will still be able to recycle the old checkbook joke. If there is any calendar, I bet it will be the Gregorian. I wouldn't be surprised, in fact, if the calendar outlasts the various religions that gave rise to it. Gregory's calendar may be around long after the last pope has been forgotten.

HOURS, MINUTES, SECONDS

Dissecting the day

Out of the right fob hung a great silver chain, with a wonderful kind of engine at the bottom . . . He put this engine to our ears, which made an incessant noise like that of a watermill, and we conjecture it is either some unknown animal, or the god that he worships; but we are more inclined to the latter opinion, because he assures us . . . that he seldom does anything without consulting it: he called it his oracle, and said it pointed out the time for every action of his life.

— Jonathan Swift, *Gulliver's Travels*

As I approach the bottom of a long, sloping driveway just off Massachusetts Avenue, a few kilometers northwest of downtown Washington, I see that I have caught the attention of an armed guard. He advances briskly from his booth and asks in a rather serious manner if he can help me. Numerous security cameras seem to be pointed in my general direction. "I have an appointment with Dr. Matsakis up at the observatory," I explain. It turns out that I'm at the wrong driveway; I have to walk around to the left, farther up the hill. Apparently a lot of visitors to the United States Naval Observatory make the same mistake.

That first driveway leads to the vice president's residence. Having no immediate business with Dick Cheney, I continue up the road. (No doubt my appearance gave me away. I suspect Mr. Cheney's visitors arrive in shiny black cars, and not on foot from the bus stop.) I wonder if the VP or his boss has ever looked through one of the observatory's many telescopes. Abraham Lincoln did. He's said to have enjoyed the view of the moon and Arcturus through the observatory's large refracting telescope.

A short walk brings me to the stately administrative building at the heart of the observatory complex. A white telescope dome rises from the west end of the building; on the roof beside it is the gold-colored "time ball" that still drops at noon each day so that ships on the Potomac can set their chronometers. The signal was no doubt more valued in 1845, when the service began, than it is in today's era of ubiquitous digital watches, radio time signals, and GPS locators.

The Time Factory

A few moments later I'm sitting down with Demetrios Matsakis, head of the observatory's Time Service department. He is dressed to the nines – gray jacket, white shirt, striped tie; if his jaw were a bit wider, he would bear a resemblance to actor Ricardo Montalbán. Matsakis has a Ph.D. in physics and worked as a radio astronomer at the observatory before being, as he puts it, "seduced by the timekeeping art." (That would seem to explain the posters of both Einstein and Stonehenge on his office wall.) He was asked to head the department about ten years ago.

When I suggest that he is the man in charge of the nation's time, Matsakis corrects me: he's in charge of the Department of Defence's time. But for most purposes, that's the same thing; many civilian applications ultimately derive their time from the USNO's clocks. Take the GPS network, for example. The satellites that make up the network depend on a set of precisely timed signals relayed between the clocks on board the satellites and the master clock here at the observatory. If the clocks are off by just a billionth of a second – one nanosecond – the system will give positions that are off by thirty centimeters.* "If you want to know where your car is, that's not too bad," says Matsakis. "But ten-nanosecond errors, twenty-nanosecond errors – the errors get bigger, and it scales right up. So that's a problem." He's not exaggerating. If you're landing the space shuttle – or even a packed 767 at O'Hare – every meter counts. Nanoseconds matter.

The master clock may be the most important machine on the site, but Matsakis explains there are nearly 100 clocks altogether. (Probably 100 on the nose "if we count the sundial," he says.) Most of them are cesium beam tube clocks: they keep time by counting the oscillations of cesium atoms, which have a natural frequency of 9,192,631,770 cycles per second. The others are called hydrogen maser clocks. They work by injecting atoms of hydrogen into a chamber called a "resonance cavity," where they oscillate with their own particular (and very stable) frequency. The technology behind the maser clocks is newer than that of the cesium clocks, and they are even more accurate. (If you want to buy one, I recommend the cesium variety. Matsakis tells me they sell for a mere $60,000 each. The maser clocks carry a $250,000 price tag.)

The challenge, however, is keeping all of these clocks in step. Each of them "has their own opinion of the time," Matsakis says. They can disagree by as much as a few nanoseconds. As the old saying goes, "A man with a watch knows what time it is. A man with two watches is never sure." Part of Matsakis's job is to develop the computer algorithms that allow the output of all those clocks to be combined into one signal that can be sent to the master clock. "The question 'What's the right algorithm?' is a subject in itself," he says. There are entire conferences devoted to working out the correct set of equations, he explains. At the time of my visit, he was organizing the Fifth International Time Scale Algorithm Symposium,

---

* Light travels at about 300,000 kilometers per second. In one nanosecond it travels one billionth of that distance – 30 centimeters, or approximately one foot.

which was held in Spain in early 2008. (I can't help wondering if the time-obsessed scientists paused for siesta along with the locals.)

Matsakis takes me on a tour of the USNO's various clocks, scattered among several different buildings on the observatory's grounds. All are housed in climate-controlled vaults, where the temperature is regulated to a tenth of a degree Celsius. The cesium clocks don't look like much. The squat, beige boxlike instruments resemble computer hard drives, or maybe high-end stereo amplifiers. The maser clocks are black, and a bit taller, about the size and shape of a hotel room's minibar.

There is quite a hierarchy of clocks at the observatory. "We have one 'Master Clock,' with a capital M and a capital C," Matsakis explains. "And we have other 'master clocks,' with lowercase m's and c's, that control their own measurement systems, and do things in parallel. They're here in case the main one breaks."

I have to ask: Has the Master Clock ever broken down?

"Oh, sure," he says. "In my ten years, it's happened twice. It seems to like to break when I'm just about to leave town. One time I was in a plane on a runway. Another time I was driving off to my son's wedding." It's not such a big deal, he assures me. "It's equipment; it can break. We have established procedures for what to do. The whole team gets together." I try to imagine a dejected, pouting "MC" being taken offline, and a bubbly, enthusiastic "mc" stepping up to the plate to fulfill the nation's precision-timekeeping needs.

The Master Clock itself – I observe it from the hallway, through a window – looks like a very ordinary stack of blue, black, and gray electronic components. Dubbed NAV-18, it has a variety of knobs and buttons; a half-dozen coaxial cables, which connect it to other machinery; and several LED displays, two of which, for some reason, read "0." The third display, with a little effort, can be deciphered as Coordinated Universal Time (UTC), which replaced Greenwich Mean Time as the world's time reference in 1972.

The Master Clock's modest appearance is deceiving: this remarkable machine "talks" to the USNO's other clocks, and constantly corrects itself to best reflect their collective timekeeping output. And it does so with staggering precision. It keeps time to within one hundred picoseconds – one hundred *trillionths* of a second – over the course of each day, every day. Had it been set when the dinosaurs went extinct, 60 million years ago, it would have gained or lost no more than about two seconds.

I glance at my watch. One of us is off by fifty seconds. I'm guessing it's me.

Keeping Time by Sun, Sand, and Water

Matsakis's clocks are the culmination of thousands of years of ever-more-precise time reckoning. We have seen how ancient skywatchers learned to track the months by following the phases of the moon; tallying the sun's daily rising and setting must be an equally primitive instinct. But dividing the day itself into smaller parts is more difficult, and marks a more recent development in our history.

Every day, the sun rises in the east, climbs high in the southern sky, and sets in the west.* Someone long ago must have noticed that if you plant a stick vertically into the ground, it casts a shadow, and the motion of the shadow mirrors the motion of the sun across the sky – and so the idea of the sundial was born. That stick would evolve into the sundial's pointer, called a *gnomon*, from a Greek word meaning "to show" or "indicate." The first sundials were probably made in the middle of the fourth millennium B.C., likely in Egypt or the Near East.

The Egyptians built sundials on both large and small scales, from giant obelisks to tiny hand-held versions. The portable "shadow clock," dating from around 1500 B.C., was an ingenious yet simple affair: when the crossbar on the T-shaped device is pointed toward the sun, its shadow falls on the ruled perpendicular shaft, allowing the user to estimate the hour. The Egyptians divided the day into twenty-four hours – an idea that

---

* At least for observers in the northern hemisphere. For those in the southern hemisphere, the sun still moves from east to west, but in an arc across the *northern* sky.

may have originated with the Babylonians.* The Egyptians used their sundials to track twelve hours of daylight, with twelve more hours assigned to the darkness of night. (The number twelve was surely considered to be special, as it is roughly the number of lunar cycles in the year.)

The twenty-four-hour day is a tradition that we've kept, but with one important change: the length of the Egyptian hour would have varied with the seasons; an hour in summer was longer because the day itself was longer. Today we use hours of a fixed length – and so we count quite a few more hours of daylight in summer than in winter.

By the time of the Roman Empire, sundials had become both sophisticated and common. In the first century B.C., an architect named Vitruvius could list thirteen kinds of sundials. They were placed in public squares and private courtyards. They soon became integral to Roman society, allowing people to structure their days, which could now be mapped out into hours, and the hours cut into halves and quarters.

Not everyone was pleased. "The gods confound the man who first found out how to distinguish the hours," the Roman playwright Plautus lamented in the second century B.C. "Confound him, too, who in this place set up a sundial, to cut and hack my days so wretchedly into small portions!"

There were other timekeeping devices that didn't rely on the sun. Sand clocks work pretty much like the modern hourglass: they run for a certain interval of time, after which one turns them over and the process begins again. One could also use a slow-burning candle, with regularly spaced notches marked on its side.

Another ancient timekeeping tool was the clepsydra, or water clock. A clepsydra can be as simple as a bucket with a small hole at the bottom from which water drips at a regular rate. Marks on the side of the bucket

---

* The Babylonians used a sexagesimal counting system – a place-value system like ours, but with a base of 60 rather than 10. (60 is convenient because it divides evenly by so many other numbers: 2, 3, 4, 5, 6, 10, 12, 20, and 30.) Our divisions of the day into 24 hours ($2 \times 12 = 24$), of hours into 60 minutes, and of minutes into 60 seconds, reflects the influence of the sexagesimal system, as does our division of a circle into 360 degrees ($6 \times 60 = 360$).

indicate intervals of time. Alternatively, water can drip into a second vessel, with its own markings to track the hours. In the law courts of ancient Rome, water clocks were used to regulate the period that each lawyer could speak. If people wanted to hear more, they would yell out *"aquam dare"* – "add water." The Roman expression for wasting time was *aquam perdere* – literally, "to lose water."

Although water clocks were used throughout the ancient world, it was in the Far East that the most sophisticated devices were built. In fact, China's mechanical water clocks predate Europe's first mechanical clocks by several centuries. The best-known and most elaborate was the so-called Heavenly Clockwork, built by a civil servant named Su Sung and begun in A.D. 1077. This remarkable machine used flowing water to turn a giant wheel at a precise, controlled rate. The wheel carried thirty-six buckets, which filled and emptied in steady succession. By the time it was completed in 1090, the clock was nearly ten meters tall and employed dozens of wheels, bells, and gongs; it was housed in a pagoda five stories high. (Sadly, when a new emperor came to power in 1094, Su Sung's clock, along with other reminders of the old regime, was dismantled and eventually forgotten. When European clocks arrived in China centuries later, they were welcomed as a "new" invention.)

All of these instruments have obvious shortcomings: sundials are useless at night and on cloudy days, sand clocks and water clocks require constant maintenance, and water can freeze in cold weather.

Church Time

In medieval Europe, the greatest demand for reliable timekeeping came from the Church. Great cathedrals and monasteries were being built across the continent, and the monks who lived within their walls followed a strict regimen of daily activities, of which scheduled prayer was the most important. Worship began with matins in the early morning and ended with vespers in late evening. (The English word for "midday," *noon*, comes from the Latin name for the midday prayer, *none*.) At night, one of the monks would have to stay awake, keeping an eye on the water clock or the sand clock. At the appointed hour, it would be his job to ring a bell, waking the others in time for morning prayer (a ritual alluded to in the children's nursery rhyme *Frère Jacques*). One can only imagine the trouble this poor monk would have been in if he fell asleep on the job.

A solution came sometime in the thirteenth century. We don't know where the discovery was made, or who should get the credit – probably a craftsman or ironsmith working somewhere in northern Europe. He may have heard stories about the magnificent water clocks used by the Chinese, or it may have been an independent discovery. Somehow, he made the breakthrough that led to a new kind of machine for measuring time.

The crucial invention is known as the "escapement." An escapement is a device that regulates what would otherwise be a continuous movement, such as the rotation of a wheel pulled by a falling weight. At regular intervals, the escapement engages, and then disengages, the rotating wheel. This slows the wheel down and, more crucially, makes it rotate at a uniform rate. The wheel, in turn, can be geared to a mechanism for striking a bell at a particular hour. "The tick-tock of the clock's escapement," as historian Daniel Boorstin has put it, "would become the voice of time."

And while the length of an "hour" read off a sundial changed with the seasons, the hour measured by a mechanical clock remained fixed. An hour in summer and an hour in winter were now the same. The creation of "equal hours," says Boorstin, was one of the great revolutions in human experience: "Here was man's declaration of independence from the sun, new proof of his mastery over himself and his surroundings. Only later would it be revealed that he had accomplished this mastery by putting himself under the dominion of a machine with imperious demands all its own." Timekeeping had become divorced from the motion of the heavens – or nearly so. At this stage, the sundial remained the most reliable timekeeper; a clock would have to be reset from time to time, perhaps daily, as it drifted from the "true" hour given by the sundial.

These early clocks ticked because of the regular motion of the escapement, but at first they had no "hands"; it was only the bell that indicated the time. Our word *clock* comes from the French word for "bell," *cloche* (German *Glocke*, Middle English *clok*) – although the same word could just as easily apply to a water clock or sand clock. Similarly, the Latin word *horologium* could apply to any kind of timekeeping device.

Although we can't say who invented the mechanical clock, or precisely when, we can be sure such clocks existed in the final decades of the thirteenth century. Records show that a fully automated, weight-driven clock was installed at Dunstable Priory in Bedfordshire, England, in 1283. Within a few decades, most cathedrals and monasteries – at least those that could afford it – likely had such a device.

The Iron Voice of Salisbury

To say that the small English city of Salisbury is charming is like saying the pyramids are old or the Great Wall of China is long. Its medieval streets are lined with wonderfully preserved half-timbered houses, and the market square bustles on Tuesday and Saturday mornings as it has for seven hundred years. Thirsty visitors in need of a pint can duck into any number of creaky, low-ceilinged pubs with typical English names (the Old Ale House, the King's Head), very English names (the Coach and Horses, the Wig and Quill), or uniquely English names (the Haunch of Venison). But the real attraction in Salisbury is the magnificent thirteenth-century cathedral and the tranquil green space – the "close" – that surrounds it. In his *Notes from a Small Island*, writer Bill Bryson declares, "There is no doubt in my mind that Salisbury Cathedral is the single most beautiful structure in England and the Close around it the most beautiful space." No wonder John Constable set up his easel across the river to capture the serene grandeur of the cathedral and the surrounding meadows.

Salisbury Cathedral boasts many records: its towering spire, at 123 meters, is the tallest in Britain; its close is the largest. Its chapter house contains one of just four surviving original copies of the Magna Carta. It also houses what is very likely the world's oldest working clock. The machine was built in the late 1300s and was originally housed in the cathedral's bell tower; when the tower was demolished in the eighteenth century, the medieval clock was put in storage and forgotten. It was re-discovered in the early twentieth century and, after some refurbishment, was placed at ground level in the north aisle, not far from the cathedral's grand western entrance, where it stands today.

On a recent visit, I met with John Plaister, whose title is as English as the city: he's the Clock-keeper to the Dean and Chapter of Salisbury Cathedral. As we listened to the clock's hypnotic tick-tocking, Plaister explained the significance of the machine's many parts. The most visible components are the vertical, geared wheels with their iron teeth. As Plaister points out, the gears turn by the power of gravity: two stone weights hang from pulleys behind the clock; as the weights fall, they pull on ropes that unwind from a pair of horizontal wooden cylinders, turning two of the geared wheels. (One of them controls the so-called time-train; the other, the clock's striking mechanism.) When the weights reach the floor, the clock has to be "wound": the weights are hoisted back up, using a pair of iron wheels that resemble a car's steering wheel.

The most important part of the clock – probably unnoticed by most visitors – is the escapement itself. It consists of two critical components: a vertical shaft known as a "verge," and a horizontal iron bar that swings back and forth, called a "foliot." Two small weights that hang from the ends of the foliot control the rate at which the clock ticks. "Of course, a clock working for more than six hundred years has had to have some replacements," Plaister says with a plummy West Country accent. "But the majority of the clock is original."

I'm struck by how little the machine resembles what we think of as a "clock." Like all such early devices, it has no hands and indeed no "face"; only the bell (now removed) would have announced the hour. The entire mechanism is housed in a cube-shaped iron frame, a bit more than a meter on a side; one can peer straight through it.

The Salisbury clock, like all mechanical clocks of that era, was not particularly accurate, easily gaining or losing as much as fifteen minutes over the course of a day. A good Roman sundial could have done the job just as well – on sunny days, at least. "When this clock was made, an hour was only divided into four equal quarters," Plaister says. "Minutes hadn't arrived yet." Not like today, where "we all tear about ... with our modern watches, to the second," he says. "They tore about to the nearest quarter or the nearest hour."

Plaister explains that although the clock's exact origins are unknown, the cathedral's records show that a man was paid to wind the clock as early as 1386. "I've no fear to say it's the oldest working clock in England," he says. "I wouldn't like to put my hand on my heart and say it couldn't be challenged, somewhere in the world. But I think it's got a good head start."

Just sixty kilometers to the west of Salisbury, about halfway between Shepton Mallet and Cheddar, sits the tiny cathedral city of Wells. It too has a medieval clock; in fact, its clock may have been designed by the same man who built the Salisbury machine. And yet the two clocks could hardly be more different.

The Wells clock has a magnificent face, with elaborate paintings of the earth, moon, sun, and stars – a colorful fourteenth-century model of the known cosmos (and one of the few surviving renderings of a geocentric – that is, pre-Copernican – world view). The clock has an hour hand that moves along a twenty-four-hour dial; the minute hand is believed to have been added in the sixteenth century.

Along with the time of day, the Wells clock displays the day of the lunar month and the phase of the moon. It also puts on a show: above the dials, a miniature medieval tournament springs to life every quarter-hour, with four tiny knights on horseback galloping in a circle – two in each direction. As the warriors spin around, one of the knights always knocks his opponent flat on his back, only to see him propped up again a few seconds later, ready for another confrontation. Over the centuries he's been knocked down some 53 million times. "We always say he should have learned to duck by now," observes Frances Neale, the cathedral's archivist. "But he never has."

These clocks were revolutionary for their day – but more advances were in store. True "striking" clocks – ringing the hours from one through twelve – were in place by the middle of the fourteenth century. For the first time in history, anyone within earshot could know the time. Clockmakers began to give their machines dials and hands – actually just one hand, at first, to show the hour; that was all that the accuracy of these early machines justified.

But clocks were becoming ever more reliable, more accurate, and more sophisticated. A few of them were dazzling showpieces that would have been a focal point for an entire city, such as the great astronomical clocks that appeared in Strasbourg, Prague, Copenhagen, and other European centers.

Before long, clocks were ringing out from courthouses and town halls, and later from banks and businesses. Wealthy citizens began to keep clocks in their homes. Miniaturization came quickly: clocks powered by tightly wound springs rather than falling weights allowed for greater portability; the first pocket watches appeared in the early 1500s. Less than a century later, Elizabeth I wore a watch on her finger. (It even had a tiny alarm – a small prong would extend and scratch her finger at the appointed hour.) Clock time was beginning to be ubiquitous.

The Value of Time

Much has been written on the role of the clock in ushering in a new, more hurried way of life. Certainly the proliferation of public clocks made time more visible; at the very least, people must have begun to think of time as something that kept passing – something that unfolded relentlessly as hour followed hour. And yet the pace of life was surely on the rise even in medieval times, before the first mechanical clocks appeared on the scene. The new technology may have simply been the latest means of satisfying a well-established urge. The development of mechanical clocks "was more a consequence than a cause of the sense of urgency that arose in the Middle Ages and Renaissance," argues historian Sara Schechner. "Clocks were seen as tools to aid the administration of civic life and their bells were used to synchronize work schedules, but the same roles were given to sundials, sandglasses, and calendars."

Indeed, the quantization of time may have been part of a larger trend of assigning numbers to previously uncounted (or poorly counted) entities – what historian Alfred Crosby has called the "quantitative revolution." Anthropologist Anthony Aveni points out that perspective painting, double-entry bookkeeping, polyphonic music, monetary standards, and a new precision in weights and measures all appeared on the scene at roughly the same time. "In a relatively brief span of years around 1300," he writes, "virtually everything in the Western world became an essence to which a number could be assigned – a sea change in the very perception of reality."

Time – now a measurable quantity – was beginning to be seen as a commodity to be valued. Although the phrase "time is money" had yet to be popularized by Benjamin Franklin, time was already being treated as such.* Wasting time was seen as not only foolish but sinful. A seventeenth-century Puritan clergyman named Richard Baxter put it this way: "To redeem time is to see that we cast none of it away in vain, but use every minute of it as a most precious thing ... Consider also how unrecoverable Time is when it's past. Take it now or it's lost forever. All the men on earth, with all their power, and all their wit, are not able to recall one minute that is gone."

The 1600s brought a new kind of timekeeper – a clock regulated by a swinging pendulum rather than a verge and foliot. One of the first to consider such a mechanism was the Italian astronomer and mathematician Galileo Galilei (1564–1642). Galileo had noticed the regularity of a swinging pendulum – he may have been inspired by the sight of a gently swaying lamp hanging from the ceiling of the cathedral in Pisa – and he even drew up plans for a pendulum clock in the final years of his life. But the first such clocks were built in the Netherlands in the 1650s, based on a design by the Dutch astronomer Christiaan Huygens (1629–95).

---

* The phrase appeared in Franklin's *Advice to a Young Tradesman* (1748). However, the sentiment is much older. In ancient Greece, an orator named Antiphon declared that "the most costly outlay is time."

Galileo showed how a pendulum could be used for precision timekeeping. This diagram was likely drawn by his son, Vincenzio, *c.* 1641. *(Science Photo Library)*

By the end of the century, the accuracy of pendulum clocks had improved dramatically, with a typical daily error falling from fifteen minutes to as little as fifteen seconds. By now, most clocks and watches had a minute hand, and the second hand would soon follow. Clocks had finally become more useful than sundials. It may not be a coincidence that, by the 1660s, the word "punctual" had come into common usage.

Trouble at Sea: The Longitude Problem

One of the greatest pressures for accurate timekeeping came from ship captains. Successful navigation required an accurate measure of both latitude and longitude, the two coordinates that mark any given location. Latitude, the distance north or south of the equator, can be read off a sextant – for example, by measuring the height of Polaris (the North Star) above the horizon. But there was no easy way to determine one's longitude, the distance traveled east or west. Countries like Britain depended on shipping and trade, and mistakes at sea were costing ships, cargo – and hundreds of lives. The problem attracted some of Europe's brightest minds, as geographers, astronomers, and craftsmen struggled to find an answer. It was also the prime motivation behind the founding of the Royal Observatory at Greenwich in 1675. (Charles II appointed John Flamsteed as the first Astronomer Royal, charging him to "apply himself with the most exact care and diligence to the rectifying of the tables of the motions of the heavens, and the places of the fixed stars, so as to find out the so much desired longitude of places for the perfecting of the art of navigation.")

In principle, longitude can be calculated – *if* you know the time difference between your current location and your home port. Suppose, for example, that you set sail from London, heading through the English Channel and out into the Atlantic. A few days later, you know you're several hundred kilometers west of England – but exactly how far have you traveled? If you knew the time back at London, then – with a little math – you can work out your longitude. Let's say you know that it's 1:00 p.m. back home in London. But from tracking the sun as it crosses the sky, you can see that your own local time is 12:00 noon. In twenty-four hours, the earth turns through 360 degrees of longitude. That means that in one hour, the earth turns through 15 degrees. Turning that around, each one-hour time difference corresponds to 15 degrees of longitude. So you must be located 15 degrees west of London.

The problem, then, boils down to knowing the time back at the home port. A good pendulum clock, set prior to departure, would do the trick, but a pendulum clock would be useless on the rolling deck of a ship. And the various portable clocks and watches that were in use at that time were hopelessly inaccurate. What was needed was a time-keeping device that was both accurate and portable – one that could cope with changes in temperature, and keep working even on a rough sea voyage.

The quest for a method of determining longitude at sea became a top priority for governments of the day. In 1714, the British parliament, through the newly established Board of Longitude, offered a prize of £20,000 – equivalent to more than $10 million today – to anyone who could solve the longitude problem.

One man who took the challenge seriously was an Englishman named John Harrison (1693–1776). Born in Yorkshire, with no formal education, Harrison devoted his life to the problem of precision time-keeping.* His four great machines – timepieces that he labored over for decades – now form the centerpiece of the gallery that bears his name in the museum at the Greenwich Observatory, just outside London.

The first three timepieces – known as H1, H2, and H3 – are hardly what we would consider "portable"; each is roughly the size of a car's engine. Weighing in at thirty-five kilograms, H1 is intimidating in its complexity. In front, it has an oval-shaped brass plate, with four large dials displaying the time. Behind that, the clock's inner workings are exposed – dozens of brass wheels and gears, and what seems like hundreds of shiny brass rods of various lengths, sticking out of the machine at odd angles. The clock still functions: its many parts slide back and forth, or spin around, as the spring-driven mechanism hums along. At sea, it kept time to about ten seconds per day – not bad, but Harrison believed he could do better.

Harrison was sure that H2 – just as big and complex – was flawed; he never even tested it. Instead he put his hopes in H3, which he believed would be his masterpiece. The third machine is an astonishing achievement. It has 753 separate parts. One of them, called a "bi-metallic strip," was a crucial breakthrough: by using a thin strip of brass attached to a

---

* Harrison's quest is detailed in Dava Sobel's popular book *Longitude* (1995).

parallel, thin strip of steel, the clock's escapement could compensate perfectly for changes in temperature. H3 was slightly smaller than H1 and H2; even so, it stood 60 centimeters high and weighed nearly 30 kilograms. For nearly two decades, Harrison built and rebuilt this third machine. But he could never get it quite right.

"Poor old Harrison spent nineteen years of his life trying to persuade it to keep stable time," says Jonathan Betts, Curator of Horology at the museum. "And he never achieved satisfaction. In later years he could only describe it through gritted teeth as 'my curious third machine.' It must have been a great disappointment to him."

Harrison's frustration may be reflected in the words he inscribed on each of his machines. The inscription on H2, engraved in large, ornate letters, reads, "Made for His Majesty George ii." The frontispiece of H3 says simply, "John Harrison."

With H3 seeming to be a dead end, Harrison tried a radically new approach. He drew up plans for a small pocket watch, and commissioned a friend to build it for him. He wanted to use it to test the accuracy of his larger clocks. Most watches of Harrison's day were just about useless as precision timekeepers. But as Harrison reflected on the design of the new pocket watch, he began to think that an accurate watch could, in fact, be made.

Harrison had finally started to think small – and it paid off. The result is the remarkable machine known as H4. The difference between it and its predecessors is like night and day. While his first three clocks were enormous, H4 is just twelve centimeters across – about the size of a dessert plate – and weighs only one and a half kilograms. In its polished silver case, it looks like a slightly larger version of an ordinary pocket watch. But H4 was anything but ordinary. With its jeweled pivots crafted from rubies and diamonds, the watch's innards were virtually frictionless. During sea trials, it lost only five seconds over two and a half months. It was the best watch the world had ever seen.

In spite of the success of H4, the Board of Longitude waffled; as Betts puts it, they kept "moving the goalposts" for awarding the prize, and gave Harrison only a series of small payments. Only after he appealed

John Harrison's ultra-precise marine chronometer, known simply as "H4." (© *National Maritime Museum, Greenwich, London*)

directly to the new king, George III, was he finally paid in full.* He died three years later.

Harrison's story is a remarkable one, even if he is only just beginning to get the recognition he deserves. "Harrison is the father of the precision watch," says Betts. "It was only after H4 that people began to recognize that we can have accurate time in our pockets and on our wrists."

In the final decades of the eighteenth century, Britain was in the midst of a sweeping transformation – we now call it the Industrial Revolution – and much of that change would have been unthinkable without precision timekeeping.

---

* In the end, Harrison was paid by Parliament, at the king's insistence. The Board of Longitude never actually awarded its prize to anyone.

Steam power – mainly derived from coal – led to a boom in manu-facturing. People began to work in factories; goods were soon moving by steamship and locomotive. By the late nineteenth century, the telegraph and the telephone carried information across vast distances in an instant. All of these developments led to a more scheduled world. Even in the early years of this transformation, the new tools of industry were chang-ing the way people went about their lives. The role of the clock was becoming paramount. As historian of science G.J. Whitrow writes,

> Steam power was the driving force of the industrial revolution. Although the old cottage-based handloom weavers often had to work very hard for a living, they could at least work when they liked, but factory workers had to work whenever the steam power was on. This compelled people to be punctual, not just to the hour but to the minute.

The steam engine may be the symbol of this period of profound change, but it was the clock that made it all possible. As historian Lewis Mumford observed, "The clock, not the steam engine, is the key machine of the modern industrial age."

The effect on society was profound, and was felt in the workplace most of all. The clock now declared, unambiguously, how much time one owed to one's employer and how much time was truly one's own. By the start of the twentieth century, work and leisure were thoroughly separated.

Trouble with the Trains: The Time Zone Problem

Time, however, remained a *local* affair. In the early 1800s, different cities kept different local times. In England, even though clocks and watches were commonplace, it was still the sun that regulated their use: it was noon when the sun reached its highest point in the sky, and this, naturally, varied from one town to the next. Even if the difference between neigh-boring towns was negligible, it added up. There was nearly a half-hour difference between Dover in the east and Penzance in the west; a twenty-minute difference separated London and Bristol. In North America, the differences were vastly greater. When it was noon in Chicago, for example, it was 12:30 in Pittsburgh, 12:55 in New York, and 1:08 in Boston.

In the age of the horse-drawn stagecoach, when it took several days to travel from one city to another, that time difference hardly mattered. But with the growth of the railways in the 1800s, people began to move much faster, and the world began to appear smaller. The German writer and poet Heinrich Heine, writing from Paris, mused on his shrinking continent in 1843:

> What changes must now occur, in our way of looking at things, in our notions! . . . Space is killed by the railways, and we are left with time alone . . . Now you can travel to Orléans in four and a half

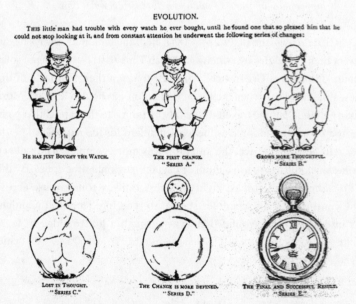

EVOLUTION.

THIS little man had trouble with every watch he ever bought, until he found one that so pleased him that he could not stop looking at it, and from constant attention he underwent the following series of changes:

HE HAS JUST BOUGHT THE WATCH.

THE FIRST CHANGE. "SERIES A."

GROWS MORE THOUGHTFUL. "SERIES B."

LOST IN THOUGHT. "SERIES C."

THE CHANGE IS MORE DEFINED. "SERIES D."

THE FINAL AND SUCCESSFUL RESULT. "SERIES E."

Man and machine gradually merge in this advertisement for the Waterbury Watch Company, 1883. *(Library of Congress)* hours, and it takes no longer to get to Rouen. Just imagine what will happen when the lines to Belgium and Germany are connected up with their railways! I feel as if the mountains and forests of all countries were advancing on Paris. Even now, I can smell the German linden trees; the North Sea's breakers are rolling against my door.

Because of the problem of time, however, achieving this "connectedness" would not be easy. The variation among local times was beginning to cause significant confusion. An explanatory note in a British timetable for the Great Western Railway, from 1841, is typical:

> *London time is kept at all stations on the railway, which is about 4 minutes earlier than Reading time; 5½ minutes before Stevenson time; 7½ minutes before Cirencester time; 8 minutes before Chippenham time; and 14 minutes before Bridgewater time.*

Britain took the first step: in 1847, all of the nation's railways began to set their schedules by Greenwich Mean Time (GMT) – the "mean solar time" observed at Greenwich.* Many Britons got their first taste of the new system in 1851, when the Great Exhibition was held in London. More than 6 million people traveled – mainly by train – to the capital that year. In 1880, GMT was adopted as the legal standard for the nation.

In North America, the situation was more complex. The railway companies started to use a number of small, regional time zones. By the 1870s, there were at least eighty of these railway zones in use across the continent. Train schedules started to read like technical manuals, as inter-city trains wound their way through a hodgepodge of local time zones.

---

* A sundial gives "solar time." What we strive to duplicate with our clocks is "mean solar time" – the time that a sundial *would* show if the sun moved at a uniform rate across the sky at all times throughout the year.

An engineer named Sandford Fleming (1827–1915) had the logical answer. Born in Scotland and raised in Canada, Fleming proposed a system for standardizing time around the globe. In 1879, he suggested that the world should be divided into twenty-four equal zones of Standard Time, each spanning fifteen degrees of longitude. Clocks in each zone would keep the same time – namely, mean solar time at the central meridian of the zone. Each zone would be exactly one hour ahead of, or behind, the neighboring zone. Critics dismissed the idea as utopian, but Fleming persisted, promoting the plan at one conference after another, year after year.

For the system to work, there would have to be a "starting point" for measuring time – a Prime Meridian from which the other zones could be referred. Naturally, each country wanted the honor of having the Prime Meridian on its own soil. Britain, however, had a clear head start: the observatory at Greenwich housed the most sophisticated telescopes and clocks, and already served as the time standard for most of the world's shipping. After much debate, the international community recognized Greenwich as the Prime Meridian. Fleming's proposal for Standard Time was adopted by delegates to the International Prime Meridian Conference in Washington in 1884.* (The French nonetheless managed to avoid mentioning Greenwich in any of their official documents. In 1898, French time was officially defined as "Paris mean time, retarded by nine minutes and twenty-one seconds" – which, as writer Clark Blaise has put it, just happens to be "identical to that of a certain leafy London suburb.")

We now take Standard Time for granted: when we fly into a new time zone, we adjust our watches by however many hours the pilot tells us to. In general, we remain oblivious to the chaos that would prevail without such a system. But the implications of Standard Time are profound. It has even played a role in forging national identities, according

---

* It is worth mentioning that the new system was a Western creation. At the conference, there was only one representative from Africa and two from Asia (from Japan and Turkey – which were also the only non-Christian nations involved).

to historian Michael O'Malley of George Mason University in Virginia, and the author of *Keeping Watch: A History of American Time*. One result of Standard Time, O'Malley told me, is a kind of "vertical bond" connecting towns and cities that are thousands of kilometers apart – but that just happen to lie, more or less, on a north-south line. "It puts me in a community with people in Atlanta – people I have nothing in common with, except that we're getting up at exactly the same time," he says. "Sunrise in Atlanta is at a very different time than sunrise here." Today, a teacher in Maine, a lawyer in Baltimore, and a shop clerk in Florida all start their working day at the same time, and, if they happen to be fans of Conan O'Brien or David Letterman, they'll switch their TVs on at exactly the same time in the evening. This kind of synchrony, O'Malley notes, is felt even more acutely during major television events such as the Super Bowl, when, during the commercial breaks – as the utility companies are well aware – a million toilets are likely to flush at the same time.

Just as Standard Time was taking hold, the machines that we use to read the time moved one step closer to us: we started to wear them. At first, wristwatches were regarded as jewelry and were worn mostly by women. That changed during the First World War, when soldiers started wearing them in the trenches. We have been wearing these intimate timepieces ever since.*

In the late 1920s, the invention of the quartz oscillator allowed for precise timekeeping beyond the capability of even the best mechanical clocks. Scientists found that when certain kinds of crystals are subjected to an electric charge, they vibrate – and the frequency of the vibration

---

* A few years ago I would have written that "many of us feel naked without a watch on," or words to that effect – but that may no longer be true. There may now be a trend toward dispensing with the wristwatch altogether, as it is rendered redundant by the LCD clocks on our cell phones. Indeed, it is now the cell phone that many of us feel naked without.

can be controlled by adjusting the size of the crystal. Those vibrations can in turn be used to control an integrated circuit – essentially a series of tiny electric switches – which can power a display, either analog or digital. (In the latter variety, there are no moving parts.) The first reliable quartz clocks were developed in the 1940s. The best of them could keep time to 1/10,000 of a second per year – a staggering leap forward from even the best mechanical timekeeper. The first quartz wristwatches appeared in the late 1960s; even the dollar-store variety can easily keep time to about a second per day. The vast majority of today's clocks and watches make use of this tiny device, one of the miracles of twentieth-century engineering.

Living on Atomic Time

Another kind of oscillator, making use of the natural frequencies of vibrating atoms, proved even more accurate. The first atomic clock was built in 1948, using molecules of ammonia. A few years later, scientists discovered how to use atoms of cesium to build an even more effective timekeeper. In a cesium clock, atoms are held in a special cavity and bombarded with microwave radiation. That radiation causes the atoms to flip back and forth between two energy states, and the rate of that flipping remains extremely stable. Cesium clocks can now be found in leading research laboratories around the world – including the U.S. Naval Observatory, where Dr. Matsakis is justifiably proud of his flock of exquisite timekeepers.

Thanks to atomic clocks, we now measure time with a precision that is actually better than that of the natural cycles that inspired them. We once used sundials to check the accuracy of our clocks, but now it is the other way around: these ultra-precise clocks can actually reveal the irregularities in the earth's rotation. ("The earth is a lousy clock," as Matsakis puts it.) As a result, in 1967, the international definition of the second was changed. Previously, it had been linked to solar time: the second was simply 1/86,400 of the mean solar day. But if the day itself is known to vary in length, that definition is flawed. Today the second is defined as the duration of 9,192,631,770 vibrations of a particular isotope of cesium.

New technologies hold the promise of even more precise time-keeping in the future. Devices known as atomic fountain clocks and ion trap clocks, developed over the last few years, are one hope. Matsakis's colleagues at the USNO are developing a fountain clock that will use atoms of rubidium. When the technology is perfected, he says, such clocks will be 50 to 100 times better than the current crop of atomic clocks. Another route is to take advantage of elements that oscillate at even higher frequencies. Strontium, for example, resonates at a frequency of 429,228,004,229,952 cycles per second (that's 429 *trillion*). Researchers at the University of Tokyo have recently claimed to be able to construct an "optical lattice clock" using strontium that can keep time with an accuracy of 1 part in 10^{18} – that is, one part in 1,000 trillion. Such a clock would be accurate to 1 second in 30 billion years.

Atomic time is determined not by any one clock or any one observatory, but rather by a global network of such clocks. Labs from around the world feed the signals from their atomic clocks to an office at the International Bureau of Weights and Measures just outside Paris, where computers work out a kind of weighted average of those times. (The USNO's contribution makes up about 40 per cent of that average.) The result, as mentioned, is Coordinated Universal Time.

But that is not the end of the story. Our clocks have also revealed that the earth itself is slowing down, spinning ever-so-slightly slower from year to year. The effect is due to tidal friction, a friction-like force resulting from the unequal pulling of the moon (and to a lesser extent the sun) on the earth and its oceans, as though a giant brake were being applied to our spinning planet. (In fact, the day is getting longer by just a bit less than one second, on average, per year. At the time of the dinosaurs, a "day" was probably around twenty-three hours long.) Left to their own devices, our atomic clocks will eventually disagree significantly with solar time. The solution is to add, from time to time, a "leap second." Leap seconds are inserted into UTC at the end of June or the end of December as needed, to keep atomic time in sync with the earth's rotation (that is, in sync with mean solar time). Without these corrections, the sun would be overhead at midnight rather than noon after a few thousand years. Since their introduction in 1972, leap seconds have been added twenty-three times.

Not everyone is happy with the leap-second system. There is always a chance that someone will make a mistake – a computer programmer will forget to implement the leap second (or input it incorrectly), and crucial clocks will go awry. What difference could a second make? In today's world, quite a bit. As we've seen, GPS systems depend on signals from atomic clocks; at the latitude of Washington, Matsakis says, a one-second error would lead to a position that's off by a thousand feet (about three hundred meters). "Do you want to have a plane crash because it thinks it's a thousand feet off from where it is?" Matsakis asks.

What is the alternative to leap seconds? One could simply try to live without them – that is, to let atomic time and solar time gradually drift apart. Before long, in that scenario, astronomers would find it hard to do their jobs: aiming your telescope in the right direction requires knowing the time with respect to the earth's rotation; if you're off by just a few seconds, the object you want to study won't even be within the telescope's field of view. Astronomers would no doubt find ways to correct for this – but as the discrepancy built up, even ordinary citizens would begin to feel the tension. Would New Yorkers really want their midwinter sun to rise only at 10 a.m.? One option would be to let the seconds accumulate until they add up to a full hour – which would take about six hundred years. Then one could add a leap hour – something we're already used to doing, thanks to Daylight Saving Time. Critics say such a move would amount to little more than shelving the problem for a few hundred years, passing the buck to the citizens of the twenty-seventh century. (The international body that decides such matters, the International Telecommunication Union, met in Geneva in 2005 to consider the issue. Their conclusion: "More time is required to build a consensus.")

The issue at the heart of such arguments is one that goes back to the Middle Ages: Do we read time *from* nature, by looking up at the position of the sun and stars? Or do we impose time *on* nature, looking down at our mechanical timepieces? So far we have always managed to find a delicate balance between these conflicting urges.

Atomic clocks are very impressive in an intellectual sort of way, and we certainly rely on them every day for more activities than we are ever likely to realize. But my favorite timepiece is still the trusty medieval clock in Salisbury Cathedral – a clock that has stood its ground for more than six hundred years, bearing witness to all the highs and lows of English history from the fourteenth century until today. When John Plaister, the clock-keeper, recounts that history, his words reveal the depth of his passion for both clock and country. "My goodness, the Black Death, civil war in England – the country split between north and south, the Wars of the Roses . . ." As he runs through the milestones in the clock's life, I think of its methodical tick-tock, tick-tock: a soft but persistent voice, oblivious to war and peace, famine and plenty, revolution and empire. For most of those centuries, the clock was up in the belfry. Perhaps it was safer up there. "It's seen Oliver Cromwell come charging about – and luckily it was too far up the tower for him to get involved," Plaister rhapsodizes. "It's seen poor Queen Elizabeth the First's ships chased 'round the English Channel. And it was ticking when the *Mary Rose* was sunk. And more recently – and more exciting to the children of England – it was ticking in this cathedral when Guy Fawkes attempted to blow Parliament up. So it has really had an incredible history." Plaister pauses. "And I wonder what its makers would have said if, at that time, we had told them that this clock would see a man on the moon. I'm sure they would have thought that we'd lost our marbles."

IN TIME'S GRASP

Time and culture

In view of all you have to do, when you waste an
hour, it seems to me a thousand . . . For I deem
naught so precious to you, both for body and soul,
as time, and methinks you value it too little.

– Letter to Francesco di Marco Datini,

a prosperous Italian merchant, from his wife, written in 1399

It is often said that our lives are increasingly rushed and more full of
stress than ever before – so often said, in fact, that it is almost a cliché
to bring it up. But we do seem to be constantly glancing at our watches,
don't we? Even when we try not to think about time, it lurks menacingly
in the background. Many of us are paid by the hour; telephone and
Internet companies bill by the minute; advertising time is sold by the
second. In the last few decades, the pace of life seems to have reached
breakneck speeds. We feel pressured to accomplish more and more in less
and less time – or so it seems. Of course, leisure time has not disappeared;
golf courses and ski resorts are doing just fine. But it seems that even
when we play, we feel rushed. Is that person next to you on the beach
checking their BlackBerry? People send us e-mails in the morning and
then phone to ask why there's been no reply by lunchtime. Having spent
the day in front of the computer, we return home – where we promptly
start checking our personal e-mail – and some of us work during the
commute. (One recent study found that even making a sandwich "has

become too much effort" for many people, forcing them to buy more "pre-packaged and prepared food items.")

The urge to cram the maximum number of activities into the minimum amount of time, notes French cultural commentator François Tournier, makes us "prisoners of the present . . . If we do not slow down, we risk becoming alienated from our own future." We're living in a "time famine," adds Harvey Moldofsky, director of the Centre for Sleep and Chronobiology at the University of Toronto. "There isn't enough time in our waking periods to accomplish all of the expectations industrial society requires of us."

But as we saw in the previous chapter, our ancestors living just a few centuries ago would have worried little about minutes and not at all about seconds; indeed, these are relatively new concepts. Chaucer had no notion of the length of a minute; Shakespeare did, but nowhere does he mention the second. And even in Shakespeare's day, the word *hour* could be used to mean "moment" – suggesting most people got along quite well without finer divisions. (An example being the final words of the "Hail Mary" prayer: *nunc et in hora mortis nostrae* – "now and at the hour of death.")

Even today, however, not everyone is in such a hurry. The way we interact with time has varied greatly across the millennia and continues to vary enormously from one culture to another – from those that obsess over time to those that barely acknowledge the existence of past and future. In some cases, as we've seen, those differences are reflected in the observatories, clocks, and calendars that we have created. But temporal differences show up in more subtle ways as well – in the religions we practice, the rituals we follow, and even the words we speak.

We saw in Chapter 2 how the Maya of Central America viewed time as an organic, animate entity. They were hardly alone: such views can be found across the globe, from tribal regions of Africa to the highly structured civilizations of the ancient Far East.

Never enough time: Grand Central Station, New York.

China: The Fabric of Time

Perhaps no other ancient culture had as complex a view of time as China. The Chinese tracked the motions of the night sky in minute detail; oracle bones from the Shang Dynasty, from the thirteenth century B.C., describe a lunar eclipse and constitute one of the oldest surviving records of a specific astronomical event. As with the Maya, celestial events were viewed as having terrestrial influence, with the appearance of each comet, eclipse, or planetary alignment seen as a divine commentary on earthly events.

Time, for the Chinese, was in part cyclic; political dynasties were seen to rise and fall in concert with celestial cycles. The sage Confucius, in the fifth century B.C., compared the ideal ruler to the North Star, around which the entire cosmos revolves.* But superimposed on this was a deep sense of

---

* The sentiment would be echoed by Julius Caesar – or at least Shakespeare's version of the emperor – who declares, "I am constant as the northern star, / Of whose true-fix'd and resting quality / There is no fellow in the firmament."

time's continuity on both short and long scales. As we saw in the previous chapter, China developed elaborate water clocks more than a century before the first mechanical clocks appeared in Europe. Yet such temporal sophistication was coupled with what would seem to be – to a Western sensibility – a very peculiar view of causality. In ancient China, time was "woven together as if in a large fabric," says David Pankenier, an authority on Chinese history, language, and culture at Lehigh University in Pennsylvania. Events in one part of the empire were seen to impact events elsewhere, regardless of their sequence in time – just as a fabric pulled in one spot causes ripples to be felt across its breadth. At least one emperor chose to abdicate rather than to rule for more than sixty years – a feat that would have been perceived as a violation of the natural order. When a dynasty was in decline – and here again we have echoes of the Mayan world view – the malaise was felt throughout the empire. "It's the whole tenor of the moment," says Pankenier. "It's like listening to the keynote of a musical piece that's off-key." For the Chinese, as historian J.T. Fraser has put it, the goal was "temporal harmony within the person, among individuals, and between society and nature."

Hinduism and Buddhism: Escaping Time's Cycles

The religions of India and southeast Asia offer a quite different view of – or, one might say, escape from – time. In the Hindu faith, cyclic time dominates; the shortest cycle, known as the *maha yuga*, lasts for 4,320,000 years. A thousand of these make one *kalpa*, and two *kalpas* make one day in the life of Brahma, the chief Hindu deity. The one-hundred-year life of Brahma works out to about 311 trillion years. Cyclic time guarantees that everything will return to a former state; history – and perhaps time itself – is an illusion. Nothing is permanent, and even death is merely a passageway to birth and renewal.

In the Hindu faith, the goddess Kali, a consort of Shiva and perhaps the most fearsome of the Hindu deities, personifies both the creative and destructive power of time. Sculptures often show the blood-stained, four-armed goddess wearing a garland of decapitated heads. She is some-times seen as a triumphant murderer, standing over Shiva's lifeless body; in other depictions, she can be seen engaging him in sexual union. But Shiva is always brought back to life, bringing all creatures back to life with him.

The devout Hindu has a way of escaping the repetition of time's endless cycles: one can merge one's consciousness with the timeless *Kaivalya* (sometimes translated as "solitariness" or "detachment"). Time is ominous and oppressive – yet through enlightenment, it can be tamed. Buddhism offers a similar path: through meditation, the Buddhist can escape the endless cycle of death and rebirth, ultimately reaching emanci-pation from time in a condition known as *Nirvāna*. (Indeed, one sect, the Madhyamikas, endorses a kind of transcendentalism that denies the reality of time altogether.) Philosopher Philip Novak has put it succinctly: "One plunges into time's terrible surf only to emerge riding its wake, awakened."

Africa: "Event Time"

In some cultures, the organic nature of time is reflected in the relative importance of events and sequences rather than marked by the clock or calendar. In many parts of Africa, "event time" prevails over "clock time."

In the 1930s, British anthropologist E.E. Evans-Pritchard made a detailed study of the semi-nomadic Nuer people of southern Sudan. He found that the Nuer do not have anything corresponding to hours or minutes; indeed, they do not have any one word corresponding to our abstracted notion of "time." Days are neither named nor numbered. As a result, they do not speak of time as an entity unto itself; it is not a thing that can be saved or wasted. Instead, time is associated with activity. The Nuer divide the year into wet and dry months; the seasons are recognized by changes in rainfall and wind patterns. Ask a Nuer in May what time of the year it is, and he'll say that "the old people are returning to the villages." Ask again in January and you'll be told that "everyone is returning to the dry-season camps." As Anthony Aveni writes in *Empires of Time*, "The temporal logic seems to be: if I'm going to church this must be Sunday; or since people are on the move going between camp and village, then it must be [the month of] *dwat* . . . Activity supersedes time in the sense that we know it." (Try to imagine a Wall Street brokerage that begins the trading day "when the employees have begun trading," or a train bound for Chicago that departs "when the people are on board.")

Nuer life is built around social structure, and while individual men and women would not necessarily know their own age, they would be able to tell you which "age set" they belong to (youth, adolescent, and so on, through old age). And although there is no time – as we imagine it – to "flow," people do indeed move though the age sets as the years pass. Because of the importance of social relationships, as Evans-Pritchard noted, the Nuer are more concerned with the past than the future, "since relationships must be explained in terms of the past."

The Mursi people of East Africa are also caught up in "event time" – but with a twist. They do keep track of the phases of the moon, which tells them which *bergu*, or month, they are currently in. As with the Nuer, each month is associated with a particular occurrence or activity: in Bergu 1 the river begins to subside; in Bergu 2 it's time to clear the riverside gardens; in Bergu 3 the sorghum must be planted; and so on. However, there is no universal agreement within the tribe as to which *bergu* is currently underway. If you ask a Mursi tribesman which month it is right now, notes Aveni, "he will probably tell you that some people in his village recently told him it's the 5th, while others said it's the 6th"; arguments over which is correct can apparently become quite heated. While anthropologists at first read this lack of agreement to suggest that the Mursi were not interested in accurate timekeeping, they later discovered it had a kind of built-in logic. In fact, if every Mursi agreed about which *bergu* it was, they would run into trouble. Because the *bergus* are linked (however loosely) with seasonal activity, the twelve *bergus* must, somehow, be reconciled with the solar year – which, as we saw in

Chapter 2, is longer than twelve lunar months but shorter than thirteen. And so, eventually, months have to be either doubled up or skipped altogether – which is more readily accommodated when one is not particularly committed to the status of the current *bergu*. "[For] these frequently mobile people," notes Aveni, "keeping time is an interactive process, a dialogue among many people based on social rules that advocate an agreement to disagree."

An interesting parallel can be seen in the "week" observed by the Umeda people, who live not in Africa but in central New Guinea. The Umeda do not track months, observes anthropologist Alfred Gell, nor do they know how many months are in a year. The seasons are divided, very roughly, as wet and dry. The Umeda lack weekly markets or a day of rest. However, Gell explains, they do count days: they measure out seven distinct days *relative to "today."* The Umeda can refer to a particular day as

> the day before the day before yesterday
> the day before yesterday
> yesterday
> today
> tomorrow
> the day after tomorrow
> the day after the day after tomorrow

"In the Umeda 'week,'" observes Gell, "today, so to speak, is always Wednesday."

Gell makes another observation that highlights the difference between the Umeda conception of time and our own. As mentioned, the Umeda do not track the lunar month, and, indeed, they are not aware that the time from one new moon to the next is a constant (roughly 29½ days). "As far as [the Umeda] are concerned the moon is like a tuber, growing in a garden, and tubers can grow quickly or slowly for unknown reasons," Gell notes. "Consequently, when the Umedas notice the waxing moon, they comment on it favourably, as if the swollen moon were a contingent piece of horticultural good fortune, not an absolutely regular and predictable astronomical event."

When time is marked by events rather than clocks and calendars, the order of those events can take on great significance. For the Luo people of western Kenya, time is literally felt to stand still unless certain events take place in their prescribed order. "The role played by the first wife, in a polygamous Luo household, is very, very clear," says Chap Kusimba, an archeologist and an authority on tribal African culture at the Field Museum in Chicago. "The first wife must be the first one to plant her crops, to prepare her farm, before all the other women. She must be the first one to weed, the first to eat," he says. Similarly, the first-born son "must be the one to marry first, whether he likes it or not . . . Things are done in a particular way. You can't break that rule." Failure to observe these rules could upset the natural order, leading, the Luo believed, to infertility or death.

Philosopher and anthropologist John Mbiti, in his book *African Religions and Philosophy*, notes that many African cultures envision "a long *past*, a *present*, and virtually no *future*"; many of the East African languages that he studied have "no concrete words or expressions to convey the idea of a distant future."* Many tribes focus on present and past occurrences because these are "real," Mbiti says; future events are not considered because, not yet having occurred, they cannot constitute time. "*Actual time* is therefore what is present and what is past. It moves 'backward' rather than 'forward'; and people set their minds not on future things, but chiefly on what has taken place."

---

* Mbiti's views have received some criticism. A.A. Ayoade, for example, says this view "is not remotely true of the Yorubas whose total perspective of time-future even extends beyond the end of this life to an after life," and adds that, as farmers, they are constantly planning for harvests and the storage of food.

Indigenous Americans: The Shadow of Time

In some cultures, time – as we think of it in the West – barely seems to exist. For many indigenous cultures in the Americas and Australia, and in a number of African and Pacific Island societies, there is no single word for "time." Evan T. Pritchard, a descendant of the Mi'kmaq Nation of eastern Canada, spent many years observing the elders of his tribe. The Mi'kmaq are quite conscious of the regular events that one would associate with time's passage: they have words for day, night, sunrise, sunset, youth, adulthood, and old age – but not for time itself, Pritchard notes in his aptly titled book *No Word for Time*. "There is no concept of time outside its embodiments in the things of nature," he writes.

Among indigenous American conceptions of time, perhaps none has triggered as much speculation as that of the Hopi of the U.S. Southwest. Early ethnographers made much of the supposed disregard for time displayed by the Hopi. American linguist Benjamin Lee Whorf, who studied their language and culture extensively in the 1930s, concluded that it "contains no words, grammatical forms, constructions, or expressions that refer to time or any of its aspects."

Instead, according to Whorf, the Hopi divide entities into two broad categories – one for dealing with physical objects and all that is taken in by the senses, the other encompassing the mental and the spiritual. What we would call "past" and "present" they put in the former group; the future, in the latter. They have plenty of verbs, but, Whorf claimed, no tenses. More recent studies have questioned that interpretation. Alfred Gell, for example, argues that the Hopi use of modalities – terms that indicate the speaker's attitude toward the content of an assertion – can do the work of European-style tenses. It also appears that the Hopi often use spatial metaphors to convey temporal facts. "Hopi is not a timeless or a tenseless language," he concludes.

Historian G.J. Whitrow agrees that it is overly simplistic to contend, as Whorf appears to, that the Hopi lived without a conception of time. Whitrow reminds us that the Hopi "have successfully developed an agricultural and ceremonial calendar, based on astronomical lore, that is sufficiently precise for particular festivals seldom to fall more than two days from the norm." Indeed, the Hopi, along with the neighboring Zuni people, were expert sun-watchers, using "horizon calendars" to track the position of sunrise from season to season – vital for planting and harvesting crops at the right time.

Backs to the Future

When politicians talk about "the way forward" in Iraq, we know that they're talking about the future and not the past. When the Temptations sing "Don't Look Back," we know we're being asked not to look backward in *time*. Those seem like straightforward metaphors, and one might guess that they would be universal – but they are not. Consider the Aymara people of South America, who have a unique way of associating space with time. In most cultures, people think of the past as behind them and the future as being out in front; this way of thinking is also reflected in the gestures and body language we use when referring to past and future. But anthropologist Rafael Núñez of the University of California in San Diego, who videotaped dozens of conversations with adult Aymara speakers in northern Chile, found the exact opposite: the Aymara point forward when they talk about the past, and gesture toward the rear when discussing the future. This reversal is also reflected in their language. The Aymara word for "front" (*nayra*, which can mean "eye," "front," or "sight") is also used to refer to the past. Meanwhile, the basic word for "back" (*qhipa*, which can also mean "behind") is used to refer to the future.

It is not clear how this way of thinking evolved. However, Núñez notes that the Aymara make a distinction between statements that they have a personal knowledge of – events that they witnessed first-hand – and things they only know second-hand. This emphasis on "what can be seen" could account for associating the past – where events are known with certainty – with the forward direction, in which things can be clearly seen; the future, meanwhile, is uncertain, and thus is associated with what is behind and unseen. Núñez found that some elderly Aymara speakers refused to even discuss the future, believing that it was meaningless to talk about events that have not happened. He also notes that this cultural attribute appears to be slowly vanishing. Although Aymara is spoken by several million people, it is found only in the highlands of the central Andes, and the younger generation, who tend to be fluent in Spanish, typically gesture in the Western fashion.

Australia: The "Dreamtime"

Perhaps the most enigmatic blurring of time can be seen in the "Dreamtime" of Australian Aborigines, a golden age that is at once long ago and also eternally present. Ancestral figures who lived in the remote past are not gone, but are seen to be embodied in those living today. The social order depends on the ritual reenactment of the lives of the human and animal inhabitants of the Dreamtime, and on passing down their stories from generation to generation; life is guaranteed only through maintaining contact with the ancestral past.* While we in the West focus on "progress" and "purpose," notes anthropologist Howard Morphy, "Aboriginal concepts of time are essentially atelic (purposeless). Or, rather, the purpose lies in the Dreaming, which is in many respects infinite and timeless."

Sociologist Mike Donaldson notes that in the Dreamtime, "time, place and people were as one. One knew the time by the place one was in, and by the company one shared." And just because a society had no word for time, Donaldson notes, does not mean that there was no

---

* This may seem alien to Western secular life, but perhaps not so alien to Western theology. Consider the celebration of the Eucharist, in which Christians reenact key aspects of the "Lord's Supper" – the last meal shared by Jesus and his disciples. As one scholar has put it, this "act of Christian worship, the saving efficacy of an event that took place outside the walls of Jerusalem in about the year A.D. 29, is ritually perpetuated, so that it is made daily available to the faithful who join in its performance. Indeed, the whole action of the rite is to make constantly and ubiquitously present an act that is definitely located in space and time."

timekeeping. "Daily time was marked by daybreak, sunrise, morning, midday, afternoon, late afternoon, sunset, evening, and night. Time could be and was counted by sleeps, moons, phases of the moon and by seasons."

It may seem that perceptions of time are as numerous as there are cultures on our planet. And yet, as different as they may appear, it is likely that there are more similarities than differences. As Alfred Gell has put it, "There is no fairyland where people experience time in a way that is markedly unlike the way in which we do ourselves, where there is no past, present and future, where time stands still, or chases its own tail, or swings back and forth like a pendulum."

The Greeks: A Diversity of Times

In ancient Greece we find not one but many approaches to the question of time, and in the writings of the Greek thinkers we can see the roots of the deepest problems that philosophers wrestle with to this day.

These descriptions emerge slowly, beginning with the epic poems of Homer and Hesiod. Indeed, Hesiod's *Works and Days*, which reads something like a farmer's almanac circa 700 B.C., is founded implicitly on the notion of time (although the word for "time" is never used); it is full of instructions on when to plant and when to harvest, with explicit references to the constellations. Homer's view is more problematic: cycles of time are evident throughout *The Odyssey* – the start of each day is heralded by the coming of "rosy-fingered dawn" – yet the hero Odysseus goes on a twenty-year voyage and does not appear to get any older during that time. (His wife, Penelope, left behind in Ithaca, similarly appears not to age. However, his son Telemachus does grow up during the interim; his faithful dog, Argos, likewise ages, and dies contented upon his master's return.)

Centuries later, the pre-Socratic philosophers began to dissect time itself. Parmenides of Elea (*c.* 520–430 B.C.), one of the most profound thinkers of his age, saw the past and future as illusions; the real world, he insisted, is timeless and unchanging. That vision stands in sharp contrast to the views of Heraclitus of Ephesus (*c.* 535–475 B.C.), who saw the universe as inherently dynamic, caught up in an endless process of creation, destruction, and change. This would become one of the key philosophical battles in the Western perception of time, and we will return to it often in the chapters ahead.

Plato (428–347 B.C.) saw a connection between time and the universe itself; in the *Timaeus*, he describes how time emerges from eternity. God, he reasons, is eternal, and seeks to make the universe eternal – but must settle for an image or reflection of eternity. The creator, he says, "resolved to have a moving image of eternity, and when he set in order the heaven, he made this image eternal but moving according to number, while eternity itself rests in unity; and this image we call time." Plato thinks of time as a kind of reflection of the rotation of the "heavenly spheres" that hold the sun, moon, and stars. This notion – that time is a characteristic of the universe rather than an independent entity – continues to influence Western thought to this day.

Plato's pupil Aristotle (384–322 B.C.) disagreed with his teacher on many fronts, and the nature of time is no exception. (From the first paragraph of his *Physics*, we know we are in for a thorough treatment: "The best plan will be to begin by working out the difficulties connected with it, making use of the current arguments," Aristotle states. "First, does it belong to the class of things that exist or to that of things that do not exist? Then secondly, what is its nature?") After much hand-wringing, Aristotle concludes that time is rooted in motion and change; it is meaningful only with respect to the events embedded in its flow. And yet time cannot be the same as motion, he reasons, because motion is "in" the body that's moving, while time is everywhere. And so time and movement become inexorably linked. "They define each other," he states. "And we measure both the distance by the movement and the movement by the distance; for we say that the road is long, if the journey is long, and that this is long if the road is long – the time, too, is movement, and the movement, is the time."

These are powerful ideas, and, as we will see in later chapters, the insights of Plato and Aristotle continue to color the debate over time.

Just as the Greeks had multiple conceptions of time, they also had differing approaches to the vexing question of how time began. Plato famously believed in a realm of ideal mathematical forms; as he describes in the *Timaeus*, the physical world came into being when those mathematical principles found material embodiment. Aristotle – no surprise – saw things differently. He felt that Plato's description implied that the world came into existence at some specific moment in the past – an idea which, in Aristotle's view, led inevitably to contradictions. Was Plato suggesting that time itself was somehow created *in time*?* This is an awkward proposition, to say the least. If one speaks of a beginning of time, Aristotle reasoned, one could immediately ask, what came before? Aristotle preferred to think of the world as eternal and unchanging, with no beginning in the past and no end in the future.

There were others, meanwhile, who embraced a cyclic view of time similar to that found in Eastern religions. The movement known as Stoicism began in Athens, under the philosopher Zeno of Citium, in the third century B.C. The Stoics believed in a "great cycle" (or "great year") involving the positions of the planets in the night sky. When the planets returned to the same relative positions, cosmic history would begin anew (an idea the Stoics may have picked up from the Babylonians). Commenting on this view in the fourth century A.D., Nemesius, Bishop of Emesa, declared, "Socrates and Plato and each individual man will live again, with the same friends and fellow citizens. They will go through the same experiences and

---

* Other interpretations suggest that Plato's view of time's origin was actually quite similar to Aristotle's. Consider this line from the *Timaeus*: "Time, then, and the heaven came into being at the same instant in order that, having been created together, if there was to be a dissolution of them, they might be dissolved together."

the same activities. Every city, village, and field will be restored, just as it was. And this restoration of the universe will take place not once, but over and over again – indeed to all eternity without end."

The idea of eternal recurrence was well established among the later Greeks. In the sixth century A.D., the philosopher Simplicius explained how the Pythagoreans – whose philosophy combined mathematics and mysticism – had embraced the idea of cyclic time. Simplicius quotes from a Pythagorean thinker named Eudemus, who has clearly thought carefully about the distinction between *similar* events recurring and *time itself* recurring:

> One might wonder whether or not the same time recurs as some say it does. Now we call things "the same" in different ways: things the same in kind plainly recur – e.g. summer and winter and the other seasons and periods; again, motions recur the same in kind – for the sun completes the solstices and the equinoxes and the other movements. But if we are to believe the Pythagoreans and hold that things the same in number recur – that you will be sitting here and I shall talk to you, holding this stick, and so on for everything else – then it is plausible that *the same time too recurs* [emphasis added].

The idea of cyclic time seems to have influenced Aristotle, who observed that "there is a circle in all other things that have a natural movement and coming into being and passing away. This is because all other things are discriminated by time and end and begin as though conforming to a cycle; for even time itself is thought to be a circle." In his text on earth sciences, the *Meteorologica*, he is even more explicit, referring to knowledge gained and lost and gained again in endless cycles: "We must say that the same opinions have arisen among men in cycles, not once, twice, nor a few times, but infinitely often."

The Birth of Linear Time

Among the members of one particular culture in the ancient Near East, in the early part of the first millennium B.C., a markedly different view of time took hold. For the Jews, history came to be seen as a sequence of unique events – one creation, one flood, one (eventual) appearance of the Messiah – all unfolding according to God's divine plan. This was a decidedly *linear* view of time. In this view, time proceeds inexorably from past to future; it is irreversible. It also demands a definite beginning: Creation happened just once. Infinity was for God, not for space and time.

One cannot attribute the origin of this view of time exclusively to Judaism; no doubt the Jews borrowed heavily from the Babylonians, who had very similar creation myths, and parallel ideas can be found in Zoroastrianism, the religion of ancient Persia. But it was the Jewish perspective that would come to be adopted by the early Christians, and that – along with certain key elements of Greek philosophy – would eventually come to form the basis of the Western world view.

However, Christian theology did not embrace Greek philosophy as a whole; certain elements, such as the Stoic view of cyclical time, had to be rejected. St. Augustine, writing in *The City of God*, cannot believe that

> just as in this age the philosopher Plato sat in the city of Athens and in the school called the Academy teaching his pupils, so also through countless ages of the past at intervals both the same Plato and the same city and the same school and the same pupils have been repeated, as they are destined to be repeated through countless ages of the future. God forbid, I say, that we should swallow such nonsense!

The adoption of linear time had a profound and lasting influence on Western thought, and would pave the way for the idea of "progress." To the historian Lewis Mumford, linear time went hand in hand with the parallel development of the mechanical clock, which we looked at in the previous chapter. The clock, he said, "dissociated time from human events and helped create belief in an independent world of mathematically measurable sequences: the special world of science." By the close of the seventeenth century, time was seen as an abstract entity that marched forward without regard for human activity.

Guns, Germs, and "CCT"

With linear time came the urge to divide and measure time with ever more accurate clocks and calendars, as we saw earlier. Together, these two developments would come to underlie modern science and industry, and, for better or for worse, would lead to the frantic pace of today's technological world. It also seems that we are exporting that high-speed lifestyle to the rest of the planet, even if we are barely conscious of doing so. In fact, the use of clocks and calendars (specifically the Gregorian calendar) to mark time has probably penetrated even farther than the West's other well-known cultural exports, such as the English language, liberal democracy, and rock music (to name just a few). For anthropologist John Postill, "clock and calendar time" – he abbreviates it to "CCT" – is "one of the West's most successful exports." Indeed, he claims there are "no reports of successful resistance to it."*

CCT can manifest itself in many ways, often without an actual clock or calendar even being present. As several anthropologists have noted, a television can easily do the trick on its own. Widely broadcast events such as soccer's World Cup can unite people from many countries in a common (if passive) activity – but the World Cup also demands that one keep a schedule: the matches are broadcast according to CCT, and viewers who

---

* Postill says, "'chronoclasm' – the intentional destruction of clocks and other time artifacts – is a word I cannot find in my dictionary." However, it does pop up in fiction: Joseph Conrad's *The Secret Agent* (1907) involves an anarchist plot to blow up the Greenwich Observatory. (It also appears in the *Urban Dictionary*, though with a somewhat different meaning: "The temporarily frazzled mental state resulting from the discovery that the actual time or date varies greatly from what you thought it was.")

tune in late will miss the opening kick-off. One study looked at the arrival of television in a remote area of Yemen in the 1970s. The coming of TV "had an immediate impact on social life," anthropologist Najwa Adra noted. It "altered everyday work and rest patterns, as most people would stay up watching TV until the electric generators were turned off at 11 p.m.," and, as a consequence, often had trouble getting up the next morning.

Japan is a particularly interesting case, as that country not only embraced clock and calendar time, but would go on to become one of the most hurried nations on earth. (As sociologist Nishimoto Ikuko notes, "While other countries were forcibly westernized, Japan chose westernization.") The change began as the country became more heavily industrialized in the late nineteenth century. In 1873, a textbook for schoolchildren included lessons on telling time by the clock. Even the language was affected: a new word, *jikan* ("time" or "hour") was introduced, largely replacing *toki* (used to refer to time as marked by Japan's traditional lunar calendar). Japanese people also began to speak of the *fun* (minute) and *byo* (second). There was no looking back. Today in Japan, a train is counted as "late" if it reaches its destination more than 1 minute behind schedule. The equivalent measure in England is 10 minutes; in France, 14 minutes; in Italy, 15. The desire to maintain a schedule can have deadly consequences: in 2005, the national railway's attempt to make up for a 90-second delay led to an accident that claimed 107 lives. As a report in the *New York Times* noted, "anywhere else in the world, a train running 90 seconds late would perhaps be considered on time."

The Pace of Life

As widespread as Western clock and calendar time has become, there is still a remarkable variation in the pace of life from one culture to another – as any traveler will have noticed. When anthropologist Kevin Birth titled his book *Any Time Is Trinidad Time* (1999), he was merely incorporating a phrase commonly heard on the island. To show up an hour late in Brazil is no big deal – but being just ten minutes late in New York, Frankfurt, or Tokyo requires an explanation.

Probably the most thorough study of such temporal variation was carried out by Robert Levine, a social psychologist at California State University in Fresno; the results of his decades-long research efforts can be found in his engrossing book *A Geography of Time* (1997). In one phase of his research, Levine and his colleagues used three different measures to gage the pace of life in thirty-one countries: the speed that pedestrians walked on city streets, the accuracy of public clocks, and the length of time it took for postal clerks to fulfill a basic request for stamps. The top five fast-paced countries turned out to be Switzerland, Ireland, Germany, Japan, and Italy; at the bottom of the list were Syria, El Salvador, Brazil, Indonesia, and Mexico. (The U.S. ranked sixteenth; Canada, seventeenth.) One can't help noticing that the "fast" countries have relatively strong economies while the "slow" countries are relatively poor. Levine does not probe this correlation in detail, but he notes, "Economic variables and the tempo tend to be mutually reinforcing; they come in a package."

In a similar study of thirty-six American cities, Levine found (perhaps surprisingly) Boston to be the fastest and Los Angeles (again, perhaps surprisingly) to be the slowest. New York was third (after Buffalo), and was also one of only two cities in the world – the other was Budapest – where the experimenters reported being personally insulted by the clerks.

Levine explains how he first became aware of the sharp differences in temporal life while teaching in Brazil. Students would show up as much as an hour late for class – and then stay just as long when the class was over. Public clocks (and people's watches) often disagreed wildly as to what time it was, and people routinely made multiple appointments for the same time – often failing to keep any of them.

In India, Levine spends hours in a lineup at a train station; the locals, who are used to such endless queuing, invite him to join them in their picnic lunch on the floor of the station. When he finally gets to the ticket window, he is told that the train is sold out, but he manages to get a ticket through *backsheesh* (bribery). In the end, "the train left late and arrived even later, none of which mattered, because the gentleman I was scheduled to meet was even later than me." In Nepal, Levine spends four days waiting for the clerks in the Kathmandu telephone office to connect his overseas call. Everyone else waits for similar lengths of time – and no one seems to mind.

Levine is careful not to pass judgment on slow-paced cultures, and cautions others not to do so: "When we attribute a Brazilian's tardiness to irresponsibility, or a Moroccan's shifting of attention to their lack of focus, we are being both careless and ethnocentrically narrow-minded." But he also notes that, at least in the U.S., many minority groups distinguish their own relaxed pace from that of the more hurried Anglo-American majority. "American Indians like to speak of 'living on Indian time,'" he notes. "Mexican-Americans differentiate between *hora inglesa* – which refers to the actual time on the clock – and *hora mexicana* – which treats the time on the clock considerably more casually."* In Mexico, Levine runs into difficulty translating a questionnaire into Spanish. He wants to ask people when they would "expect" someone to arrive for a certain appointment, when they "hoped" that person would arrive, and how long they would "wait" for them to arrive – only to discover that in Spanish the same verb, *esperar*, is used for all three.

In the few remaining hunter-gatherer societies, Levine notes, the pace of life is perhaps the most relaxed to be found on our planet. The Kapauku people, who live in the western highlands of New Guinea, do

---

* While it is apparently okay for anthropologists to describe these cultural variations, a recent incident in Canada showed that careless remarks about such differences can cause offense – especially when aimed at an individual. In 2007, an Inuit artist named Jonas Faber was wondering why he was being audited by the Canada Revenue Agency, and requested his government tax file through the Freedom of Information Act. He was shocked to see the auditor's comments, which included the statement "As is typical of natives, he doesn't have the same sense of urgency as we would have in complying with a deadline." Faber called the comments "dehumanizing" and filed a complaint with the Canadian Human Rights Commission, which has begun an investigation.

not believe in working two consecutive days, while the !Kung bushmen of southern Africa usually work about two and a half days per week, usually for six hours a day, he says. Levine notes the stereotype of life being slower in warmer nations: all of the slowest countries in his thirty-one-country survey have tropical climates.

Our ancestors began measuring time out of necessity: agriculture demanded attention to the seasons, and reaping the largest harvests demanded a knowledge of the sun and stars and their endlessly repeating motions. But time's grasp would turn out to reach much farther, influencing every facet of public and private life. For some, time was organic and flexible; for others, it ran in cycles, mirroring the cycles seen in nature. For still others – including ourselves – time is inanimate, linear, and relentless.

There are signs that some of us have had enough. As writer Carl Honoré elaborates in his recent book *In Praise of Slow* (2004), a variety of "slow" movements have begun to pop up across Europe and North America – Slow Food, Slow Cities, and, yes, Slow Sex. The manifesto of the Slow Food movement, which originated in Italy, begins, "Our century, which began and has developed under the insignia of the industrial revolution, first invented the machine and then took it as its life model. We are enslaved by speed and have all succumbed to the same insidious virus: Fast Life." (Honoré decided to write the book after he came across an article about a series of classic fairy tales that had been condensed into one-minute versions for the benefit of harried parents. He thought it was great – and then asked himself, "Have I gone completely insane?")

Fast living may seem firmly entrenched, but Western culture is not static, and the way we perceive time continues to evolve. As journalist Kate Zernicke has pointed out, the introduction of the cell phone – which became ubiquitous in industrialized nations toward the end of the 1990s – is making time "squishier than ever." "Squishy" here does not mean slow; it just means that we are now more "connected" than ever, and this, in turn, is radically changing the way we manage our time. As rushed as our culture may be, Zernicke says, it contains pockets of "soft time" – especially when friends plan their evening and weekend activities. Many of us live in "a bubble in which expectations of where and when to meet shift constantly because people expect others to be constantly reachable," she says. "Eight-thirty is still 8 o'clock as long as your voice arrives on time – or even a few minutes after – to advise that you will not be wherever you are supposed to be at the appointed hour." (All those who are guilty of this behavior, please raise your hands.) One study, Zernicke says, concluded that "the United States has become more like Brazil, where time has been spongy for generations." In Brazil, on the other hand, "people who used to just arrive late now complain that they have to call and explain."

Is the Western penchant for cramming more and more experiences into each moment of time in any way connected to the great variety of non-Western traditions that we've surveyed? Perhaps. The endless cycles of Eastern religions, and the promise of an escape from time through enlightenment, are one strategy for confronting the inevitability of death. Quite possibly our own desire to live every moment to the fullest is driven, in part, by a nagging suspicion that, in today's secular world, "you only live once."

THE PERSISTENCE OF MEMORY

A bridge across time

> Memory's vices are also its virtues, elements of a bridge across time which allows us to link the mind with the world.
>
> – Daniel Schacter

> To think is to forget. . .
>
> – Jorge Luis Borges

How is it that the three-pound mass of soft gray matter inside our heads is able to perceive the passage of time? More than three hundred years ago, the English scientist Robert Hooke, a contemporary of Newton, pondered this very problem:

> I would query by what sense it is that we come to be informed of Time; for all the information we have from the senses are momentary, and only last during the impressions made by the object. There is therefore yet wanting a sense to apprehend Time; for such a Notion we have; and yet no one of our Senses, nor all together can furnish us with it.

There is still no simple answer to Hooke's question, although there has been endless educated guesswork and, particularly in the last few decades, enthusiastic scientific inquiry.

Biologists have long known that the human body responds to the rhythms of our natural environment (indeed, plants and other animals do so as well); the burgeoning field of "chronobiology" has sprung up to examine the body's responses to these rhythms in health and in sickness. The best-known rhythms are those that rise and fall on a daily cycle: our heart rates, metabolism, digestion, and many other functions appear to be synchronized with the natural cycle of day and night. These are the so-called circadian rhythms, from the Latin words for "about" (*circa*) and "day" (*diem*).* Numerous other cycles follow shorter or longer scales.

And then there is the brain itself – the organ that endows us with conscious experience. We know the brain contains some 100 billion nerve cells, or neurons – by coincidence, that's about the same quantity as the number of stars in a typical galaxy – and we know that each of those neurons can make some 10,000 synaptic connections with other neurons. And from those connections the brain does – well, everything. Francis Crick, co-discoverer of the structure of DNA, called this the "astonishing hypothesis": the fact that this sea of neuronal activity endows us with our sense of self, our awareness of the world, our consciousness. And that would include, presumably, our awareness of the passage of time.

Psychologists and neuroscientists have long wondered if there is something like an "internal clock" within the brain – some kind of mechanism that allows it to keep track of time. But they are now beginning to

---

* And most of us are aware of how easily these rhythms can be disrupted – when we have to work at night, for example, or when we experience jet lag from traversing multiple time zones.

doubt that there is any one specific brain structure that acts as such a timekeeper.* Instead, our consciousness of time seems to be distributed over many brain regions, each with its own ways of recording information about temporal distance and sequence. "Many complex human behaviors – from understanding speech to playing catch to performing music – rely on the brain's ability to accurately tell time," says Dean Buonomano of UCLA's Brain Research Institute. "Yet no one knows how the brain does it." Two other prominent psychologists, Thomas Suddendorf and Michael Corballis – we will hear more from them shortly – share that sober assessment: "How internal clocks, order codes, or other processes give rise to adult human concepts of the time dimension remains unclear."

Part of the problem is that the study of time in the brain is not only a relatively new field of investigation, but also one that cuts across disciplines, blurring the boundaries between many narrower fields. Among those subdisciplines are electrophysiology (the study of the electrical properties of cells and living tissue) and psychophysics (the study of the relationship between physical stimuli and our subjective experience of them); new techniques in brain imaging and computational modeling are also playing a role. As neurobiologist David Eagleman has said, these new fields are gradually "contributing to an emerging picture of how the brain processes, learns, and perceives time."

The latest research is motivated by a number of overarching questions: How do our brains encode and decode information as it streams in, over time? How are signals from different brain regions coordinated with one another across short intervals of time? How well do our perceptions of duration reflect the outside world? What factors influence our

---

* There are, however, specific regions associated with regulating circadian rhythms. The most important is called the "suprachiasmatic nucleus," or SCN, near the base of the brain. When the SCN is removed from rats, their circadian rhythms become impaired.

temporal judgments? "Despite its importance to behavior and perception," says Eagleman, "the neural bases of time perception remain shrouded in mystery."

While we are only gradually beginning to see *how* the brain interprets time, there is one obvious and rather broad function of the brain for which time seems to be quite central: memory. Hooke, in fact, went on to conclude that memory was the "organ" through which we perceive time:

> Considering this, I say, we shall find a Necessity of supposing some other Organ to apprehend the Impression that is made of Time. And this I conceive to be no other than that which we generally call Memory, which . . . I suppose to be as much an Organ as the Ear, Eye, or Nose, and have its Situation somewhere near the place where the nerves from the other Senses concur and meet.

Neuroscientists today do not generally speak of memory as an organ – it is more often described as a process, or a combination of processes – but their findings have begun to show just how complex and multifaceted the brain's memory systems are. The science of memory is an enormous subject, too large for one meager book chapter. Instead of attempting to summarize what researchers have learned about the workings of memory, I will focus more narrowly on those functions that can best illuminate the connections between memory and time.

Remembering the Past, Imagining the Future

Can you picture the house you grew up in? If you close your eyes, can you imagine walking through your family's living room and making your way to the kitchen? What about events from your childhood – playing with a favorite toy, or celebrating a birthday? Chances are, even if you have just an average memory, you can conjure up memories of this sort without much difficulty, and perhaps even in substantial detail. Psychologists call this "episodic memory" – a term coined by Canadian neuroscientist Endel Tulving in the early 1970s. What about looking through time in the other direction? Can you imagine some future event that you'll take part in? A trip to the grocery store tomorrow, a holiday somewhere warm next winter?

It turns out that we're very good indeed at projecting our minds across time, mentally calling up images associated with events from our past, or things that we imagine will occur in the future. (Novelists have often played with this idea: the events in Virginia Woolf's *Mrs. Dalloway*, for example, unfold in a single day; in the mind of the central character, however, that one day blossoms into many decades.) We are so good at such projection, in fact, that we can leap mentally back across the ages; you can imagine events that took place before you were born and – although this is trickier – events in the remote future, long after you're gone.

Psychologists and cognitive scientists call this remarkable ability *mental time travel*.* It is defined, roughly, as the combination of episodic memory and the ability to anticipate future events. Without it, there would be no planning, no building, no culture; without an imagined picture of the future, our civilization would not exist. There is no question that humans are quite good at mental time travel – but there is a great deal more that we would like to understand. Can other animals do it? If not, when did it arise in hominid evolution, and why? At what stage does it appear in childhood? What can it tell us about the way we imagine the passing of time – or of the nature of time itself?

————

* Not to be confused with physical time travel, which we will look at in detail in Chapter 8. Incidentally, when we engage in mental time travel, we typically also engage in mental *space* travel – if you grew up in another city, remembering your childhood involves mentally leaping across space as well as time. That term, however, does not seem to occur in the scientific literature.

Tulving, now eighty, has retired from the University of Toronto but carries on an active research program at the Baycrest Rotman Research Institute in the city's north end. His focus is memory, but he has not shied away from deep questions about the nature of reality. In his younger days, his interests were "more philosophical," he says, leading him "to wonder about things like time, when it began, and what was there before it existed."

Tulving feels there are many misconceptions about memory and time – the most prevalent being the perceived strength of the connection between them. We are constantly aware of time's apparent flow, and memory, one might guess, is the medium through which we encounter that flow. But Tulving's decades of research have shown that this is not the case. "Most forms of memory that exist – not only in humans, but in any animal – have nothing to do with time," he says. Most facets of memory – the ability to learn a new skill or remember a fact, for example – operate strictly within the here and now. But there is one notable exception: episodic memory, which Tulving believes is unique to humans, gives us the ability to peer back across time, using our imagination to revisit just about any event that we choose.

It is only in the last few decades that psychologists have begun to investigate the connection between episodic memory and the capacity for mental time travel, and to speculate about how human beings have come to possess them to such a sophisticated degree. Tulving, who has studied episodic memory for much of the last half-century, was one of the first scientists to use the term "mental time travel." He stresses the fact that, to the person doing the remembering – whom he calls the "rememberer" – an instance of episodic memory seems real: it is felt to be an accurate (or at least highly plausible) replay of a past event, a part of one's own unique personal history. As he put it in his book *Elements of Episodic Memory* (1983), "Remembering, for the rememberer, is mental time travel, a sort of reliving of something that happened in the past."

Some of the most interesting recent work on mental time travel has been carried out by two southern-hemisphere psychologists whom we have already briefly heard from: Thomas Suddendorf of the University of Queensland, in Australia, and Michael Corballis of the University of

Auckland, in New Zealand. Suddendorf and Corballis have argued persuasively that the capacity for mental time travel gave our ancestors an invaluable edge in the struggle for survival. They believe there is a profound link between remembering the past and imagining the future. The very act of remembering, they argue, gives one the "raw material" needed to construct plausible scenarios of future events and act accordingly. Mental time travel "provides increased behavioural flexibility to act in the present to increase future survival chances." If this argument is correct, then mental time travel into the past – remembering – "is subsidiary to our ability to imagine future scenarios." Tulving agrees: "What is the benefit of knowing what has happened in the past? Why do you care? The importance is that you've learned a lesson," he says. "Perhaps the evolutionary advantage has to do with the future rather than the past."

Modern neuroscience appears to confirm that line of reasoning: as far as your brain is concerned, the act of remembering is indeed very similar to the act of imagining the future. That may seem strange at first; after all, the Red Queen in Lewis Carroll's *Through the Looking Glass*, who remembers the future instead of the past, strikes us as absurd precisely because we usually consider "memory" and "the past" as inexorably linked and as having little, if anything, to do with the future. But while we do not "remember" the future, we do picture it, and it turns out that we do so in ways that closely parallel our efforts to picture past events. Brain imaging studies have shown that, for either activity, we make use of similar regions in the brain's frontal and temporal lobes.

Harvard psychologist Daniel Schacter, writing in a recent issue of *Nature Reviews – Neuroscience*, says one can think of the brain "as a fundamentally prospective organ that is designed to use information from the past and the present to generate predictions about the future. Memory can be thought of as a tool used by the prospective brain to generate simulations of possible future events."

When I met with Schacter recently, he explained that this is still a rather novel way of interpreting the function of memory – and it could help us understand why it evolved in the first place. "We tend to think of memory as having primarily to do with the past," he says. "And maybe one reason we have it is so that we can have a warm feeling when we reminisce, and so on. But I think the thing that has been neglected is its role in allowing us to predict and simulate the future."

Imagining the future requires us to do several things. We must call on another facet of our brain's memory system, known as "semantic memory" – our general knowledge of facts about the world – to establish the framework for the imagined scene. (You can hardly imagine that a trip to France might include a stay in Paris unless you remember that Paris is in France.) However, and perhaps more importantly, it also requires us to use our episodic memory: your imagined holiday will draw on your memory of real holidays that you've had in the past, memories of real hotels, restaurants, and so on. Such memories are often flawed (as we will see shortly), but that is perhaps because their primary purpose has little to do with providing accurate recollections of the past. As Suddendorf and Corballis suggest, episodic memory "may be part of a more general toolbox that allowed us to escape from the present and develop foresight, and perhaps create a sense of personal identity." In fact, they claim, what we have traditionally thought of as memory's primary role – allowing us to conjure up the past – may be "only a design feature of our ability to conceive of the future."

Other observations support this idea. Psychologists have found that imagining an event in the remote future is more difficult than imagining an event closer at hand, just as remembering a more remote past event is more difficult than remembering an event that happened more recently. We also seem to lose both "directions" together: as we age, our powers to use episodic memory begin to fade, along with our ability to envision the future.

Most dramatically, patients who suffer from certain kinds of amnesia and have lost their ability to recall past events are equally frustrated in their efforts to imagine the future. A poignant example is a patient known to psychology as K.C., a Toronto man whom Tulving and other psychologists have studied intensively over many years. As a young man, K.C. was in a motorcycle accident. He suffered severe brain damage from a head injury and has a complete loss of episodic memory.

In many ways, K.C.'s behavior is perfectly normal. He can play chess and play the piano; these are skills he learned before the accident, ones that take advantage of a third kind of memory system, known as "procedural memory." He can walk around his immediate neighborhood without getting lost. His semantic memory and his ability to use language are unaffected. But his autobiographical memory – the story of *who he is* – is missing. He remembers no personal events from his past. He cannot say what he did yesterday – and he is equally unable to say what he may do tomorrow. His mind simply "blanks" when he tries to answer such questions. As one psychologist has put it, K.C. is "completely rooted in the present, with no ability to move backward or forward cognitively in time." The irony, of course, is that he is not aware of his predicament. Daniel Schacter has said that "you really become a shell of a person, a fragment, when you're confined to the present moment," and that being in such a state has "a massive effect on one's sense of self." And yet K.C. has rated his level of happiness at four out of five.

A San Diego man known as E.P. suffers from a nearly identical condition. Fifteen years ago, an infection destroyed large portions of his brain's temporal lobes. He has forgotten his past and cannot form new memories. Writer Joshua Foer gives a moving description of E.P. in a recent *National Geographic* cover story: "Without memory, E.P. has fallen completely out of time. He has no stream of consciousness, just droplets that immediately evaporate . . . Trapped in this limbo of an eternal present, between a past he can't remember and a future he can't contemplate, he lives a sedentary life . . . He is trapped in the ultimate existential nightmare, blind to the reality in which he lives." And yet his daughter reports that E.P. is "happy all the time. Very happy. I guess it's because he doesn't have any stress in his life."

We all place enormous value on memory. Perhaps a certain amount of forgetting is just as valuable.

Mutts © Patrick McDonnell / King Features Syndicate

The Planning of the Apes

Is the capacity for mental time travel uniquely human? Over the last few decades, a number of researchers have suggested that animals are effectively stuck in the present, with no ability to recall the past or imagine the future.* (Tulving, for example, declared, "Remembering past events is a uniquely familiar experience. It is also a uniquely human one.") More recently, studies involving the great apes, certain birds, and other animals have challenged that view, although the conclusions remain controversial.

The best-known investigations of the cognitive capabilities of animals are those that looked at the use of language among the great apes. According to Suddendorf and Corballis, however, no ape has ever communicated anything to suggest it was remembering a past event. The chimpanzee named Panzee, for example, was able to communicate the location of hidden food – it could direct a human to that location – but this "does not prove that she *remembered* the hiding event itself, just as one can know where the car keys are without remembering the event of putting them there," Suddendorf and Corballis write. The "'linguistic' outputs" of trained apes, they believe, "do not include reports of travels down memory lane. They have not provided evidence of mental time

---

* Of course, many animal species appear to plan for seasonal environmental changes: squirrels bury nuts, bears hibernate, and birds migrate. Such activity, however, is generally seen as instinctual (it cannot be "unlearned") and not as an example of mental time travel.

travel. There is so far no use of tense, nor any sense that the animals are telling stories about previous or anticipated episodes."

Chimpanzees in the wild seem to display more compelling evidence of foresight and planning. They fashion sticks into "spears" in one location and then use them for collecting termites at another location; they have also been seen to carry stones from one place to another in order to crack open nuts at the new location. But one must be cautious in interpreting such observations. The use of tools by itself does not reveal planning, cautions cognitive scientist Daniel Povinelli, director of the Cognitive Evolution Group at the University of Louisiana. Are the chimps really contemplating future events? "They're thinking," says Povinelli. "But what are they thinking about?" One crucial question is whether they're imagining a future termite-hunting or nut-cracking session – or merely working to alleviate the hunger that they feel *now*. As Suddendorf and Corballis argue, "The anticipations do not go beyond the *context* of the present," and thus should not be construed as mental time travel.

Birds, Brains, and Breakfast

Interestingly, the strongest evidence for mental time travel in animals comes not from the great apes but from a bird known as the scrub jay. Scrub jays routinely cache food in various locations and then retrieve it at a later time. Moreover, they can discriminate based on how much time has elapsed since the food was hidden: the birds will recover recently cached worms rather than nuts, because fresh worms are more palatable; but if the worms have been hidden for too long, they'll choose the nuts instead, knowing, apparently, that the worms will no longer be fresh.

In one of the most intriguing experiments involving scrub jays, psychologist Nicola Clayton and her colleagues housed the birds on alternate days in two different compartments – one in which the jays always received "breakfast," and one in which they did not. Then the birds were unexpectedly given extra food in the evening, at a location where they could access either compartment. The jays promptly cached their surplus – and they preferentially cached it in the "no breakfast" compart-ment. Because the birds were not hungry at the time of the caching, the researchers claim that the birds truly anticipated the hunger they would experience the next morning. For Clayton, now at the University of Cambridge, the implication is clear: the jays "can spontaneously plan for tomorrow without reference to their current motivational state, thereby challenging the idea that this is a uniquely human ability." (Her team also observed another intriguing behavior: it seems that scrub jays remember if they were being watched by other jays as they hid their food. If a bird is being watched, it generally returns at a later time and moves the food to a new location. Moreover, this response is strongest in birds that have themselves stolen the food of others – suggesting that it "takes a thief to know a thief.")

Psychologist William Roberts, reviewing the most recent findings in the journal *Current Biology*, concluded that both scrub jays and non-human primates "can anticipate and plan for future needs not currently experienced," and that these studies "suggest that some animals can men-tally time travel into both the past and the future." (This seems to be a concession on Roberts's part: five years earlier, he wrote that animals "may be aware of only a permanent present, whereas people readily see the world from numerous time perspectives.") Thomas Zentall, in another recent review, came to the same conclusion – that "the ability to imagine past and future events may not be uniquely human."

As impressive as the scrub jay studies are, the ever-skeptical duo of Suddendorf and Corballis say they are inconclusive. It is quite possible, they argue, that the birds "remember" where they have cached the food without mentally reconstructing the past. The researchers remain "unconvinced that

---

* Suddendorf and Corballis insist that they are not being smug about the results: "We are not making these 'kill-joy' arguments because of any preconceived notion about the way the world should be. We would be delighted if it could be established that other species have mental time travel. It would be a serious challenge to many humans' anthropocentric worldview and would have profound moral implications." But they say the evidence just isn't there. "We maintain that the data so far continue to suggest that mental time travel is unique to humans."

any of these cases shows true mental time travel."* Tulving remains similarly unconvinced. He says he is glad that Clayton stopped short of referring to the jays' ability as "episodic memory," instead referring to it as "episodic-like" memory.

The Evolution of Mental Time Travel

Even if the capacity for mental time travel is uniquely human, it is quite possible that the precursors for such an ability could have been present not only in early hominids, but even in the common ancestor of humans and the great apes.* Humans and apes are both highly social creatures, and it is possible that keeping track of what group members are up to – trying to predict, for example, what one's companions are going to do next – allows one to hone the skills needed for tracking objects across space and time. In that case, as Suddendorf and Corballis concede, such precursors to mental time travel may well be observed in today's great apes. Full-fledged mental time travel, on the other hand, may be a more recent development; Tulving, for example, estimates that it emerged more recently – perhaps in the last fifty thousand years – when *Homo sapiens* was well established.

In all likelihood, the capacity for mental time travel did not develop in isolation but rather alongside other crucial cognitive abilities. "To entertain a future event one needs some kind of imagination," Suddendorf and Corballis write, "some kind of representational space in our mind for the imaginary performance." Language could also play an important role. Our language skills embrace mental time travel by the use of tenses and recursive thinking; when we say "A year from now, he will have retired," we're imagining a future time in which some event – which has not yet happened – will lie in the past. That sentence was written in the future perfect tense; as Suddendorf and Corballis point out, this is just one of

---

* The last common ancestor of the genus *Pan* and our own genus, *Homo*, is thought to have lived between 5 and 6 million years ago.

some thirty tenses in the English language, "and reflects the close relation between language and mental time travel." (It is interesting to note that mental time travel, though rooted in our brain architecture, may still vary significantly from one culture to another. As G.J. Whitrow has observed, "although our awareness of time is a product of human evolution, our ideas of time are neither innate nor automatically learned but are intellectual constructions that result from experience and action.")

Suddendorf and Corballis suggest that mental time travel may have been "a pre-requisite to the evolution of language itself." If mental time travel is indeed unique to humans, it may help us understand why complex language is also, apparently, unique. (Another ability that may be closely related is what psychologists call "theory of mind" – the ability to discern that others have mental states that may not be identical to one's own. For example, it is only by the age of three to four that children are able to discern that another person may hold a false belief.)

Around 2 million years ago, the first members of our own genus, *Homo*, appeared on the scene, endowed by natural selection with much larger brains (in proportion to body size) than any creature that had lived before. Perhaps, Suddendorf and Corballis argue, the emergence of these larger-brained hominids reflected a selection "for such interrelated attributes as theory of mind, language, and, we propose, mental time travel." We already saw in Chapter 1 how early hominids made and transported tools for future use; to Suddendorf and Corballis, this was just the first outward sign of a blossoming new cognitive ability – one that would lead our ancestors to prosper as never before. Mental time travel, they argue, "offers the ultimate step in adaptation to the future."*

For well over a million years, there were multiple hominid species roaming the plains of Africa, Asia, and Europe at any one time. Only one – *Homo sapiens* – survived. We have already looked at the final episode

---

* Not that every realization about the future is necessarily welcome. As mentioned in Chapter 1, and as Suddendorf and Corballis acknowledge, an awareness of the future brings with it an awareness of one's own mortality. Suddendorf and Corballis admit that this "introduces new kinds of mental stress."

of that story: the disappearance of the Neanderthals some 25,000 years ago. According to Suddendorf and Corballis, our species won out "because of continued advances in foresight, language, culture and coordinated aggression, leaving us as the sole survivors of an extraordinary evolutionary arms race. We may be the only species capable of mental time travel because the others competing in our niche have become (or were made) extinct."

The result was a creature that walked upright, made use of tools, employed sophisticated language, and had, for the first time, a full-fledged conception of "past" and "future." Is this how the concept of "time" came into existence? As Suddendorf and Corballis write, "The mental reconstruction of past events and construction of future ones may have been responsible for the concept of time itself, and the understanding of a continuity between past and future."

Mental Time Travel in Children

Mental time travel does not come quickly to a child. While the jury may be out in the case of animals, there seems little doubt that children up to about eighteen months of age live in an "eternal present." Every desire, every utterance, seems to be rooted in the child's view of her immediate surroundings as they appear at the present moment. A child's sense of time develops gradually, in parallel with memory and the use of language.

Infants do show signs of short-term recollection at a very young age: at three months, a child can recognize a mobile she saw one or two weeks earlier. By the age of two, children can reproduce the order in which a series of objects was presented to them many weeks earlier.* They have

---

* Although young children can clearly form memories, those memories do not last. Most adults have no memories from before the age of three or four. Young children are simply unable to store long-term memories – probably, researchers believe, because certain key brain structures such as the hippocampus and the neocortex are not yet adequately developed.

learned to use the word "now," and later begin to use the word "soon," but have almost no vocabulary concerning the past. The word "tomorrow" is almost always learned before the word "yesterday."

It is only later, between the ages of three and five, that children can reason about – and describe – past and future events, and make plans for future activities. By around age four, children can give accurate answers to questions about events that happened yesterday and about events that are likely to happen tomorrow. (Three-year-olds can provide answers to such questions – but the answers are typically incorrect.)* As William Roberts suggests, children under four "may not have the linguistic representational skills to conceive of time as a dimension going backward and forward from the present moment." (This is also the age at which children will voluntarily choose delayed gratification – refusing a treat offered now in the knowledge that a more substantial reward will come later.)

By around age six there is no looking back. Children by this time routinely speak of yesterday, today, and tomorrow. They recall past events and imagine the future. They might start counting down the days until Christmas or a birthday weeks in advance – with very concrete expectations (and perhaps demands) about what that day will bring. A six-year-old child is cognitively light-years ahead of the smartest chimp or jay that has ever lived. Later still – between eight and ten – children begin to see time as an abstraction, "a single common time in which all events happen," as G.J. Whitrow has put it. (Interestingly, most ten-year-olds believe they age by one hour when clocks are turned ahead for Daylight Saving Time; most fifteen-year-olds realize that the time indicated by a clock is just a convention.)

At least, this is what we observe with children raised in the West. As Whitrow cautions, the way children develop their sense of time is

---

* Such experiments are fraught with difficulty. As psychologist Janie Busby points out, "it is not clear whether 3-year-olds perform poorly because they cannot yet travel mentally in time, or because they could not yet understand the questions appropriately."

deeply dependent on their cultural environment. (Just a moment ago we heard Roberts's concern that young children may not be able to "conceive of time as a dimension going backward and forward from the present moment" – but why, exactly, *should* they conceive of it this way?) In his book *Time in History* (1988), Whitrow points out that children in other cultures – he mentions a study carried out in Uganda and another among Australian Aborigines – found it "difficult to relate the time they read on the clock to the actual time of day," not because of any difference in intelligence, but because "their lives, unlike ours, are not dominated by time." A more recent assertion from the anthropologist Alfred Gell – echoing the concerns of Robert Hooke – emphasizes the difficulty we face in examining our perceptions of time: How much of that "perception" is rooted in biology, and how much is cultural? "We have no dedicated sense organ for the measurement of elapsed time," Gell writes. "To speak of the 'perception' of time is already to speak metaphorically."

The Failings of Memory

Mental time travel may indeed be the cognitive rudder that allows our brains to navigate the river of time. But the two branches of this voyage, even though they involve similar brain processes, feel quite different. When we imagine the future, we know what we picture is really just an educated guess; we may be right in the broad brushstrokes, but we are almost certainly wrong in the details. We hold memory to a higher standard. We feel – most of the time – that our memories are more than guesses, that they reflect *what really happened*. When confronted with a conflicting account of how last week's party unfolded, we cling to our beliefs: *He must be mistaken; I know what I saw*. With time, however, memories fade. We may look back to a journal entry or flip through a photo album to bring those memories back to life. For the narrator in Marcel Proust's *Remembrance of Things Past*, the taste of a *petite madeleine* – a lemon-flavored pastry – is enough to carry him back through time to the village of his childhood. Often, remembering a happy moment from long ago does seem like the next best thing to reliving that moment. But just how reliable are those memories?

It is hardly a surprise to hear that our memories often fail us. We have all had that awkward moment when, encountering someone we know, we can't quite bring to mind their name – and then either confess to our shameful failing or fake it and hope that the name somehow surfaces. And most of us have looked frantically for a lost wallet, purse, or set of keys, only to find it in some rather obvious place. A certain degree of memory loss is normal with aging, while neurological diseases such as Alzheimer's can virtually destroy memory altogether.

What may be more surprising is that, even for healthy adults, the confidence we feel in our most vivid memories is no guarantee that those memories are accurate. Often, we are dead wrong. One well-known experiment – so easy to conduct that you can try it with a friend – involves lists of words with related meanings. The experimenter reads a list of words such as *tired*, *bed*, *awake*, *dream*, *night*, *blanket*, *snore*, and *yawn* – words that are all related to a "lure" word that is not presented, in this case *sleep*. When the subjects are later given another series of words, they can easily spot new words that are unrelated to the theme words (*kitchen* or *butter*, for example). However, they also frequently claim – wrongly – that the lure word had been presented on the earlier list. Daniel Schacter says he has tested this before audiences of nearly a thousand people and that it is common for 80 to 90 per cent of them to report "hearing" the word that wasn't there. Occasionally people are so certain that they heard the word, they accuse the experimenter of lying about it!

Schacter has a theory about why we make such errors: it is more important, in evolutionary terms, that we remember the gist of events that we encounter than the specific details. The gist of a story is something we can process and react to; an endless stream of details just gets us bogged down. "The positive spin on this is that it indicates that we're very good at getting at the general sense, or gist, of what we've been exposed to," Schacter says. "The downside is that it highlights the fact that we're not so good at retaining all the details of exactly how we acquired knowledge ... So we can misattribute our strong sense of familiarity or recollection to something that never happened."

When the event in question is something that we have *all* witnessed – a famous movie, for example – we often develop a kind of collective false memory. We seem to recall that Humphrey Bogart, as Rick, says "Play it again, Sam" in *Casablanca*. (He doesn't. Rick says, "You played it for her and you can play it for me," and "If she can stand it, I can! Play it!") A nearly-as-famous line from *The Treasure of the Sierra Madre* also causes trouble. The bandit leader, played by Alfonso Bedoya, never actually says, "Badges? We don't need no stinking badges!" (The actual line is, "We don't need no badges. I don't have to show you any stinking badges.") My favorite, as a science journalist, is the recollection we seem to have that Carl Sagan, in his TV miniseries *Cosmos*, used the phrase "billions and billions" – perhaps even that he used it regularly. In fact, he swore he never did – although Johnny Carson, doing endless comedic impersonations of Sagan on *The Tonight Show*, did use the phrase quite often.

Even if our susceptibility to false memories has a silver lining – in helping us remember the gist of what happened – it is still unsettling. And lists of words and misremembered movie quotes are just the beginning. Scientists have discovered that more elaborate false memories can be instilled in subjects with relative ease, both intentionally and unintentionally. In one much-publicized study, Elizabeth Loftus and her colleagues at the University of California at Irvine were able to convince nearly one-third of their subjects that they had been lost in a shopping mall as a child. (Prior interviews with close relatives confirmed that the subjects – aged eighteen to fifty-three – had never actually had such an experience.) More controversial examples of false memories have involved recollections of alleged physical or sexual abuse that were "repressed" and later said to be recovered with the help of a therapist – often using hypnosis, "guided imagery," or other controversial practices.

Memories in a Flash

There is another kind of memory that has always seemed to stand out – and that psychologists have recently begun to scrutinize. If I asked you what you were doing on September 9 or 10, 2001, you would probably have great difficulty remembering. But, chances are, you remember exactly where you were and what you were doing on the morning of September 11 of that year. Such memories, associated with traumatic or momentous events – especially events of national significance – are known as "flash-bulb memories." That term was coined by two Harvard professors, Roger Brown and James Kulik, as they studied people's recollections of the turbulent events of the 1960s and '70s. The name is self-explanatory: the memories are so powerful that they seem imprinted on our brains like a photograph. To an earlier generation, the assassination of John F. Kennedy was such an event; more recently, the explosion of the space shuttle *Challenger* in 1986, the acquittal of O.J. Simpson in 1995, the death of Princess Diana in 1997, and of course the terrorist attacks of 2001 fit the pattern. Brown and Kulik believed that the way these memories are encoded in our brains is fundamentally different from run-of-the-mill memories. It must involve a special brain mechanism – they called it a "Now Print" – that freezes the moment when we hear of such an event.

Because of the momentous nature of the 9/11 terrorist attacks, it is not surprising that they have become the definitive case study for such traumatic public events. Even the fact that we refer to it by the date that it happened – "9/11" or "September 11th" – suggests that the date itself has already made a permanent imprint on our collective memories.*

Psychologist Elizabeth Phelps of New York University recently carried out a detailed study of 9/11 memories, including, for the first time, using brain imaging techniques to examine the mechanisms behind such recollections. I met with Professor Phelps on a recent July afternoon –

---

* Although, incredibly, a survey by the *Washington Post* in 2006 – just five years after 9/11 – found that 30 per cent of Americans could not say what *year* the attacks had happened in (95 per cent, however, could give the month and day).

one of those sweltering days when the air above New York's sidewalks is like the inside of a pizza oven, with the sun nearly overhead and the skyscrapers casting only the most feeble shadows. The eighth floor of the psychology building at NYU was, for some reason, hardly any cooler.

Phelps explains that her study turned up a number of intriguing findings. Three years after the attacks, she and her colleagues questioned people who were in downtown Manhattan – within a few kilometers of the World Trade Center – as well as those who were in Midtown, some seven or eight kilometers away. The intensity of her subjects' 9/11 memories, it turns out, depended on how close they were to the site of the attacks. Those who were nearest to the towers reported more detailed, vivid, and confident memories of that day than did those who were farther away.

Phelps believes there is a straightforward explanation: those who were close to the towers had an emotional reaction. They could see, hear, and even smell what had happened. "People that were closer were more likely to have felt threatened that day – running for cover, that sort of thing," Phelps says. "They were more likely to report actual sensory experience of the event." For those farther north, as horrific as the day's events were, there was less sense of being in physical danger. In fact, the recollections of those who were farther from the towers were not qualitatively different from memories they had of other personally significant events – moving to New York, a memorable birthday party – that they had experienced the previous summer. In other words, only the downtown group had true flashbulb memories.

This idea is supported by the results of the brain scans that Phelps and her team conducted. They used an fMRI scanner to study brain activity in the subjects as they were asked to recall various aspects of their experiences of 9/11.* Phelps found that those who were closer to the

---

* fMRI stands for "functional magnetic resonance imaging," a powerful method for measuring neural activity inside the brain. (For obvious reasons, there are no data available for the moment when the memories of 9/11 were formed.)

towers showed increased activity in an area known as the amygdalae – a pair of small almond-shaped regions deep inside the brain's temporal lobes, known to be associated with the processing of emotionally charged memories, and with the fear response in particular.

Intriguingly, while traumatic events leave us with very intense, vivid memories – just as Brown and Kulik proposed – those memories are not necessarily more accurate than those of other events. Several studies have shown that flashbulb-type memories become less accurate over time, just like ordinary memories do. In one survey, psychologists found that, just two and a half years after the *Challenger* disaster, people gave starkly different answers on a basic questionnaire compared to the answers they gave the day after the event; another study found that, after three years, one-third of respondents told inconsistent stories.

Similar results have been found in the case of the September 11 attacks. "It's hard to convince people that they don't know what they were doing on 9/11," Phelps says. But studies have found that people are "no more accurate in knowing the details of what happened on 9/11 than some party that they went to the night before," she says. "What distinguishes these memories is the *feeling* that they're accurate."

Why are we so confident in our recollection of a traumatic event when it may in fact be no more accurate than our recollection of some run-of-the-mill event? Phelps believes that it's to our advantage for our brains to ignore extraneous details. "Emotion tells you what's important," she says. "If you're trying to escape from a tiger, the details don't matter. It matters that 'It's a tiger – get the hell out of here.'" That sounds very much like what Schacter said about our difficulty with those word lists: we remember the gist, not the details, because the gist is more important. "We think maybe it's to help you make fast decisions in the face of highly emotional events," says Phelps. "That's our pet theory. We can't prove it."

Inside the Brain of Bush

If your own memories of 9/11 are beginning to fade, you're not alone. In the months after the attacks, even the President of the United States seemed to have trouble remembering the day's events. As UCLA psychologist Daniel Greenberg has noted, George W. Bush's recollection of September 11 evolved significantly during that time. On December 4, 2001 – about three months after the attacks – Bush recalled,

> I was in Florida. And my chief of staff, Andy Card – actually, I was in a classroom talking about a reading program that works. And I was sitting outside the classroom waiting to go in, and I saw an airplane hit the tower – the TV was obviously on, and I use[d] to fly myself, and I said, "There's one terrible pilot." And I said, "It must have been a horrible accident." But I was whisked off there – I didn't have much time to think about it, and I was sitting in the classroom, and Andy Card, my chief who was sitting over here, walked in and said, "A second plane has hit the tower. America's under attack."

Just over two weeks later, in an interview with the *Washington Post*, he remembered the day's events differently. As the *Post* reported,

> Bush remembers senior adviser Karl Rove bringing him the news, saying it appeared to be an accident involving a small, twin-engine plane. In fact it was American Airlines Flight 11, a Boeing 767 out of Boston Logan International Airport. Based on what he was told, Bush assumed it was an accident. "This is pilot error," the president recalled saying. "It's unbelievable that somebody would do this." Conferring with Andrew H. Card, his White House chief of staff, Bush said, "The guy must have had a heart attack." . . . At 9:05 a.m., United Airlines Flight 175, also a Boeing 767, smashed into the South Tower of the trade center. Bush was seated on a stool in the classroom when Card whispered the news: "A second plane hit the second tower. America is under attack."

By January 2002, Bush's recollection had changed once again:

> I was sitting there, and my Chief of Staff – well, first of all, when we walked into the classroom, I had seen this plane fly into the first building. There was a TV set on. And you know, I thought it was pilot error and I was amazed that anybody could make such a terrible mistake. And something was wrong with the plane or – anyway, I'm sitting there, listening to the briefing, and Andy Card came and said, "America is under attack."

The President can't seem to remember if he was told of the first crash by Karl Rove, or if he saw it on TV – and that second option is not actually possible, since the only footage of the first impact (recorded inadvertently by a TV crew working on an unrelated project) was not broadcast until much later. Not surprisingly, conspiracy theorists had a field day, claiming that if Bush had seen images of the first crash, he and his "co-conspirators" must have installed their own cameras to record the event. Greenberg asks, "Are we obliged to believe that the President is smart enough to carry out a horrific conspiracy to attack America, but dumb enough to reveal it – twice?"

Actually, a more benign explanation is available, Greenberg says, if we "consider the frailties of human memory." Part of the problem is that we are exposed to powerful images of such events in the days and weeks that follow (and never more so than for 9/11). Eventually the pictures themselves may color our memories of the original events. (In one Dutch study, carried out after the crash of an El Al cargo jet in Amsterdam in 1992, 60 per cent of those questioned said they had seen TV footage of the moment of the crash – even though no such footage exists.) We also may come to feel that TV was the *source* of news that we in fact first received by some other means – a conversation, perhaps, or a phone call. (This is an example of what psychologists call "source amnesia.") In addition, psychologists have found that we tend to combine recollections from different times – inadvertently building up an inaccurate memory of the event in question based on real memories of events that happened earlier or later (sometimes called the "wrong time slice" error).

In the case of Bush's changing recollections of 9/11, Greenberg concludes that "the President, like most Americans, presumably saw the footage many times in the subsequent months, including footage of the first crash when it became available. Then, when the President tried to remember how he heard about the first attack, he did just what so many other people have done, retrieving information from the wrong time slice and recalling a vivid, memorable image instead of the more mundane statements of Karl Rove." It may be worth noting that even as the President's recollections changed, one element remained virtually constant: that he learned about the second impact when his Chief of Staff, Andrew Card, whispered the news in his ear. It may not be a coincidence that a photograph that captured that very moment was widely published in the days and weeks after 9/11.

ISAAC'S TIME
Newton, Leibniz, and the arrow of time

Nature, and Nature's Laws lay hid in Night.
God said, Let Newton be! and all was Light.

— Alexander Pope

"Join me in singing the praises of Newton, who opens the treasure chest of hidden truth," the English astronomer Edmond Halley said in his preface to his colleague's masterwork, *Philosophiae Naturalis Principia Mathematica* (*Mathematical Principles of Natural Philosophy*). "No closer to the gods," Halley gushed, "can any mortal rise." It was well-deserved praise: with the *Principia*, Isaac Newton (1642–1727) laid out the mathematical framework that would serve as the foundation of physics for more than two hundred years.*

But the *Principia* almost never saw the light of day. Described in his youth as a "sober, silent, thinking lad," Newton lived an intensely private life. His greatest insights were scribbled by candlelight in his rooms at Cambridge; when an outbreak of the plague forced the university to shut down, he relished in the even deeper seclusion of his family's rural estate at Woolsthorpe in Lincolnshire. The latter period – an eighteen-month interval beginning in late 1665 – is sometimes called his *annus mirabilis*,

---

* A brief account of Newton's life and work – along with an overview of the Scientific Revolution – can be found in Chapter 3 of my earlier book, *Universe on a T-Shirt*.

Isaac Newton declared that "Absolute, true, and mathematical time . . . flows uniformly." For Newton, events unfold against a fixed backdrop of absolute space and time. *(© Science Museum/Science & Society)*

or miracle year. "In those days I was in the prime of my age for invention," he later recalled, "and minded mathematics and philosophy more than at any time since." In 1667, with the plague in retreat, Newton returned to Cambridge; by now his most important ideas had already crystallized in his mind. He had developed a mathematical basis for Galileo's dynamics, which he encapsulated in his three laws of motion; he had worked out the rules for a new kind of mathematics involving infinitesimal quantities, which we now call "calculus"; he had begun to investigate light and color; and, in a stunning leap of the imagination, he had shown how a falling apple and a whirling planet or moon were simply responding to the same force, which he spelled out with mathematical precision in his law of universal gravitation. He was twenty-four years old.

The Road to Principia

Yet Newton kept nearly all of these ideas to himself. Men of learning from across Europe would write to him; often he did not write back, choosing "to decline correspondencies by Letters about Mathematical & Philosophical matters," as he later recalled, finding they "tend to disputes and controversies." What little he wrote on physics – mostly unfinished treatises – would collect dust in his study in Trinity College for more than twenty years, alongside his tracts on alchemy and theology. But his attitude changed after a visit from Halley in August of 1684. Halley, based in London, knew that other scholars were on the verge of publishing ideas similar to those that Newton had been working on. (Robert Hooke, for example, appears to have thought of the inverse-square law of gravitation at about the same time as Newton.) Newton realized that if he was to be recognized for his revolutionary contributions to science, he would have to work fast. Writing at a feverish pace, he began to set down everything he knew about the structure of the universe. In doing so, he briefly retreated even further from the outside world. Often he would forget to eat; when he did opt to join his colleagues in the dining hall, he would sometimes get lost on the short walk. (On leaving his chamber, he would sometimes turn left instead of right, his assistant later recalled; on realizing his mistake, he would turn back but then once more bypass the hall and return to his room. Occasionally he would "write on his Desk standing, without giving himself the Leasure to draw a Chair to sit down in.") The *Principia* was finally printed in 1687 – and was immediately hailed as the most important treatise on physics ever written (an honor many would say it still deserves).

Time lies at the very heart of Newton's description of the world. His goal was to describe motion mathematically – and motion, of course, is a change in position over time. Yet what he actually says about time in the *Principia* has long puzzled physicists and philosophers. "I do not define time, space, place, and motion, as being well known to all," he says in the opening pages of his magnum opus – and then proceeds to do just that. (No doubt he was influenced by his former teacher, the mathematician Isaac Barrow, who once remarked, "Because mathematicians frequently make use of time, they ought to have a distinct idea of the meaning of the word, otherwise they are quacks.") Why does Newton feel he has to define "time" if we all intuitively know what it is? He says that his aim is to distinguish "mathematical time" from the "common" notion of time that we all harbor, so as to eliminate "certain prejudices." And so we come to his famous definition:

> Absolute, true, and mathematical time, in and of itself and of its own nature, without reference to anything external, flows uniformly.

What, exactly, is Newton saying here? Motion, again, is change over time – but time measured *how*? In Newton's day, clocks that kept time to better than a few minutes per day were still a novelty; anyone interested in precision timekeeping looked to the heavens and the regular motions of the sun, moon, and stars. Yet even these motions – as Newton understood better than anyone – were variable and imperfect. The example he mentions in the *Principia* is the "equation of time," the technical term for the discrepancy between the sun's daily movement across the sky – this is "solar time" as read by a sundial – and the idealized *average* of that movement, today referred to as "mean solar time." The discrepancy is no small affair: the time shown by a sundial can be as much as twenty minutes ahead or behind mean solar time. So where do we turn for flawless timekeeping? Newton realized there simply isn't a perfect clock anywhere for us to rely on, either on Earth or in the sky; even the stars, obeying the same laws as terrestrial objects, must be imprecise timekeepers. Underpinning these imperfect physical clocks, Newton reasoned, there must be a true "universal" clock: a perfect cosmic chronometer whose precise movements real clocks can only approximate.

Newton's view of time built on – but also departed from – the recent work of Galileo and Descartes. Galileo had envisioned time geometrically, as a line marked off at regular intervals; Newton's predecessor, Barrow, shared that vision.* René Descartes (1596–1650) saw time as a

---

* In fact, Barrow spoke of time in language very similar to that which Newton would later use: "Time does not imply motion, so far as its absolute and intrinsic nature is concerned; not any more than it implies rest; whether things move or are still, whether we sleep or wake, Time pursues the even tenour of its way."

measure of motion but considered the idea of duration as something sub-jective, "a mode of thinking." And although he developed a coordinate system for analyzing space geometrically, Descartes does not seem to have thought of time in geometrical terms. Newton went further by envisioning both time and space as geometrical structures that had a real existence. (As the philosopher Philip Turetzky points out, Newton may have been influenced by the growing prevalence of mechanical clocks, which, imperfect as they were, encouraged "the analogy of time with space" and reinforced "the priority of time over motion.") Newton's universal clock ticked away at a rate independent of stars and planets; independent of our perceptions. It was simply *there*. Should all the matter in the universe disappear, it would *still* be there. It was fundamental.

The Problem with "Absolute Time"

From Newton's framework we can derive all kinds of equations that let us predict how an object will move. Back in high school we learned several of the most useful ones. Suppose you have an object moving along at a steady speed. In that case, the distance that it covers will be given by the equation $d = v \times t$ ("distance equals velocity [speed] multiplied by time"). Similar (and only slightly more complicated) equations tell us how a body will move in response to a constant force, giving it an accelera-tion. Time enters into such equations as a *parameter* that describes how one quantity changes with respect to another. Newton needed a frame-work of uniform time and space in order to develop laws of this sort; it allowed him to treat time and space as abstractions. Indeed, his defini-tion of absolute time is preceded by a definition of absolute space phrased in almost exactly the same words: "Absolute space, of its own nature, without reference to anything external, always remains homogeneous and immovable." For Newton, this abstracted version of space and time was essential. Without absolute time – if we decide, for example, that speeds and distances must be measured against some particular local clock – Newton's laws lose their universality.

A number of philosophers immediately objected to Newton's idea of absolute time. They argued instead for a "relational" view of time – the idea that time only makes sense in relation to the motion of physical bodies. Among Newton's contemporaries, the German mathematician and philosopher Gottfried Leibniz (1646–1716) was the most prominent supporter of the relational view. In the year before his death, Leibniz carried on an animated exchange of letters with a supporter of Newton named Samuel Clarke (1675–1729), an English theologian. Time, the relationists argued, is simply a way of comparing one event to another. In the relational view, time is *not* independent of the material objects that make up the universe. Just the opposite, in fact: the physical objects and their motions are what *define* the passage of time. One might argue that this more closely matches our experience of the world: We do not "see" time, just as we do not see space. What we perceive are *events* in time, and objects in space.

Physicist Lee Smolin of the Perimeter Institute in Waterloo, Ontario, offers a helpful analogy. Imagine an empty concert hall in which the only sound is the ticking of a metronome that someone has left behind. The metronome is Newton's imagined absolute time, ticking

The German philosopher Gottfried Leibniz rejected Newton's "absolute" space and time. Instead he proposed a "relational" view, in which time can only be measured in relation to the motion of physical objects. *(George Bernard, Science Photo Library)*

away independently of all else. Then the musicians – perhaps a string quartet or a jazz ensemble – enter the hall. Ignoring the metronome (perhaps they can't even hear it), they begin to play. The "time" that emerges in their music, Smolin explains, is now "a relational time based on the developing real relationships among the musical thoughts and phrases." The musicians listen only to one another, and "through their musical interchange, they make a time that is unique to their place and moment in the universe." For Newton, the time of the musicians "is a shadow of the true, absolute time of the metronome." For Leibniz, on the other hand, the metronome "is a fantasy that blinds us to what is really happening"; the only time we encounter in the hall is what the performers "weave together."

It is not easy to work out a system of mechanical laws that encompass relational time; Leibniz certainly did not know how to do it. (Einstein did, as we will see in the next chapter.) And the relational view comes with its own difficulties: If time is motion, what happens when all motion ceases? Would time stop? A relationalist would have to say yes: no motion, no time. In Newton's scheme, time would march on, somehow, in the metaphysical background.

Newton vs. Leibniz

The battle between Newton and Leibniz was fought on theological grounds as well. This was an age of belief, and great thinkers like Newton and Leibniz had to do more than explain the physical world; they also had to show that it made sense for God to create the world the way he did. But Newton and Leibniz approached this issue quite differently. According to Leibniz, God does nothing on a whim; there must be a reason – a rational explanation – behind God's every action. This is called the "principle of sufficient reason." Leibniz believed that if time were absolute – continuing even when no change is observed – then time must have been passing even before God created the universe. In that case, God created the universe at some particular time. But why *that* particular time? Why not five minutes earlier or five minutes later? After all, in the Newtonian scheme, every moment is alike. (The best that Newton's ally Clarke could offer in reply was that sometimes God *did* do certain things on a whim. That did not satisfy Leibniz, who said such a notion "is plainly maintaining that God wills something, without any sufficient reason for his will.")*

In the relational picture, it isn't meaningful to speak of time in the absence of events, so time simply doesn't exist before the creation of the universe. In Leibniz's view, God didn't create the universe *in* time; rather, he created time *along with* the universe. (Mind you, Leibniz had his own troubles with time. Is time "real," if an instant has no duration? "For how can a thing exist," he asked, "whereof no part does ever exist? Nothing of time does ever exist, but instants; and an instant is not even itself a part of time.")

Newton was aware of Leibniz's objections, but he nonetheless considered his own theory to be a testament to the glory of God. Indeed, Newton's conception of absolute space and time may have been anchored to his belief in an eternal, omnipresent deity. In his published writings, that connection is seen most clearly in the "General Scholium" – a brief appendix added to the second edition of the *Principia* in 1713:**

---

* The debate continues. Philosopher J.R. Lucas, for example, suggests that God could have "chosen to set off the big bang" when he did because, ultimately, he "must at some time decide to get going." He likens it to some of his undergraduate students, who may not have wanted to get up out of bed but, "eventually, for no particular reason, they finally did decide to get up."

** Newton's voluminous theological writings, many of them unpublished, have only recently seen the light of scholarly inquiry; they show Newton every bit as obsessed with theological matters as with science, perhaps more so. Though he was a devout Christian – he owned thirty bibles at the time of his death – his religious views were wildly unorthodox. He denied the Holy Trinity, which he felt smacked of polytheism. This was a criminal offense in Newton's day; had his views been made public during his Cambridge years, he would at the very least have been expelled from Trinity College.

> He is eternal and infinite, omnipotent and omniscient, that is, he endures from eternity to eternity, and he is present from infinity to infinity ... He is not eternity and infinity, but eternal and infinite; he is not duration and space, but he endures always and is present everywhere, and by existing always and everywhere he constitutes duration and space.

As historian of science Stephen Snobelen writes, "This God of dominion is the God of Newton's faith *and* his natural philosophy. This is also the God of Newton's absolute space and time ... God comes first, and hence absolute space and time are predicates of God's infinite extension and eternal duration."

Newton also embraces the ancient "argument from design," the idea that evidence for a Creator can be seen in the natural world. He declares in the General Scholium, "This most beautiful system of the sun, planets, and comets, could only proceed from the counsel and dominion of an intelligent and powerful Being."

Interestingly, Newton would come to be seen as the father of the "clockwork universe," a cosmos that may have been set in motion by God at the beginning of time but that no longer needs divine meddling.* Newton's view, however, was just the opposite: he believed in a Creator who was constantly active, sustaining the laws of nature and intervening when necessary. For Newton, the universe and its workings were utterly contingent on the existence of God. Here, again, Leibniz disagreed: How could God – the embodiment of perfection – be so clumsy as to build a universe that needs regular maintenance?

---

* Although often associated with Newton, the clockwork metaphor is actually much older. It can be traced back to at least the thirteenth century, with the work of such thinkers as Nicolas Oresme and Johannes de Sacrobosco; later supporters include Descartes and chemist Robert Boyle (1627–91). Some philosophers would actually consider Newton an "anti-mechanical" thinker because his concept of gravity allowed for "action at a distance."

The Flow That Wasn't There

Let's turn again to Newton's definition: "Absolute, true, and mathematical time . . . flows uniformly," he tells us. But what, exactly, is flowing? The so-called flow of time is one of the most argued-about phenomena in all of philosophy. Usually when we say something flows, we mean that it flows at some particular rate with respect to something else. A river, for example, flows at some rate with respect to its banks, and we could measure that rate in liters per second or similar units. But what does time flow with respect to? And at what rate? To say that time flows by at "one second per second" tells us nothing at all. If time is a "thing" whose passage we can measure, then we must be measuring it with respect to some deeper "hypertime." In the 1960s, the philosopher Jack Smart gave a clear account of the problem:

> If time flows . . . this would be a motion with respect to a hyper-time . . . If motion in space is feet per second, at what speed is the flow of time? Seconds per what? Moreover, if passage is of the essence of time, it is presumably the essence of hypertime, too, which would lead us to postulate a hyper-hypertime and so on ad infinitum.

As the philosopher Huw Price points out, just as we can't meaningfully speak of the rate at which time flows, we also cannot meaningfully speak of the *direction* in which it flows. Saying it flows "from past to future" does not get us very far, as this is how past and future are defined.

After mentioning the flow of time in his famous definition, Newton never brings it up again. In fact, his laws of motion do not even help us distinguish past from future. All of his equations are *time-symmetric*, which means that they are equally valid as descriptions of natural phenomena no matter which way time "goes."

When we think of the power of Newton's laws, we usually think of how they help us predict future events, from the path of a projectile to the orbits of the planets. Perhaps the quintessential example is the predicted return of Halley's Comet. Halley used observations of earlier comet sightings, along with Newton's equations, to predict that a bright comet seen in 1682 would return in 1759. (Which it did, though Halley, who died in 1742, did not live to see his prediction borne out. He was eventually honored, of course, in having the comet named after him.) A book on my shelf tabulates upcoming solar and lunar eclipses for the next decade, timed to the nearest minute, and the tables could continue for the next thousand years if the author had so desired. When NASA publishes its data for the next few years worth of eclipses, the figures are given to the nearest second, and some parameters are given to the nearest tenth of a second.

Such predictions take advantage of the *deterministic* nature of Newton's laws: if you know the state of a system at some particular moment, you can, in principle, predict the state of the system at some arbitrary time in the future. The French mathematician and astronomer Pierre-Simon Laplace (1749–1827) enthusiastically embraced the deterministic world view when he declared:

> We may regard the present state of the universe as the effect of its past and the cause of its future. An intellect which at a certain moment would know all forces that set nature in motion, and all positions of all items of which nature is composed, if this intellect were also vast enough to submit these data to analysis, it would embrace in a single formula the movements of the greatest bodies of the universe and those of the tiniest atom; for such an intellect nothing would be uncertain and the future just like the past would be present before its eyes.

Notice the phrase "the future just like the past": Laplace recognized that Newton's formulas work just as well in reverse; one can use them to make predictions in either direction of time. After tracking some particular astronomical body, for example, astronomers can work out its orbital parameters and then proceed to "predict" (we might say "retrodict") where it would have been at some arbitrary time in the past. (This technique is of immense value to archeo-astronomers and historians, who can date ancient texts by working out the dates of any astronomical events that may be mentioned, such as eclipses or planetary alignments.)

But can complicated objects like people and animals "go" backward? Here we have a dilemma, because common sense says they cannot; most phenomena we observe in nature are clearly *irreversible*. And yet it is not Newton's equations that forbid such reversals; as we've seen, his equations are symmetric with respect to time. This mathematical symmetry between past and future, embodied in Newton's laws, was noted by Lord Kelvin in the journal *Nature* in 1874:

> If the motion of every particle of matter in the universe were precisely reversed at any instant, the course of nature would be simply reversed for ever after. The bursting bubble of foam at the foot of a waterfall would reunite and descend into the water; the thermal motions would reconcentrate their energy, and throw the mass up the fall in drops reforming in a close column of ascending water . . . living creatures would grow backwards, with conscious knowledge of the future, but no memory of the past, and would become again unborn.

Of course, we don't actually see the fanciful events described by Kelvin. Broken teacups do not spontaneously reassemble, scrambled eggs do not unscramble, and recorded music played backward sounds very little like "music" at all (notwithstanding the alleged tribute to "my sweet Satan" in Led Zeppelin's "Stairway to Heaven" and hints that "Paul is dead" in certain Beatles songs). Instead, time – in spite of the mathematical symmetry in Newton's laws – seems to have a definite direction.

In Search of Time's Arrow

Think of a ball rolling along the surface of a billiards table. If we imagine that the surface is frictionless and the collisions with the rubber cushions perfectly elastic, the ball will roll around forever, bouncing off the cushions every few seconds, its speed remaining constant. If we film a few minutes worth of this admittedly dull scenario, we end up with a movie that plays just as well forward as backward. Upon seeing the film, the viewer has no way of knowing in which of the two possible directions the frames of the film were recorded. (The same applies on larger scales, so long as the motions are simple enough: a film of the earth revolving around the sun also looks perfectly plausible in reverse.) In accordance with Newton's laws, the behavior of the billiard balls is time-symmetric. If we add a second ball, again the motions look the same forward and backward. They're more complicated now, because in addition to hitting the cushions, the balls sometimes hit each other. But as before, the film looks just as plausible forward as backward. If we have ten balls bouncing around the table, still there is no obvious "direction" in time – so long as they start off randomly scattered on the table.

But suppose now that we impose some order on the array of balls, by racking them up using the plastic triangle provided for just this purpose. We put the cue ball in position, take aim, and "break" the carefully ordered array of balls. From order we now have disarray. A few clumps of balls may remain, but as the next player shoots, the disorder – the randomness – increases again. Physicists use the term *entropy* to quantify the amount of disorder of a system – so we could say the entropy increases as each player takes his turn.

Now the events most definitely are *not* time-symmetric. A film shown in reverse would be immediately identified as such. We know from experience that a random array of billiard balls will not spontaneously arrange itself into a neat and compact triangular pattern. We don't see such decreases in entropy in nature.

We can see a similar pattern whether the objects involved are macroscopic, like billiard balls, or microscopic, like molecules of gas. Imagine that we have a container with two compartments, separated from each other by a valve. Suppose one compartment is filled with nitrogen gas, the other with oxygen. Now we open the valve. What happens? Immediately some of the nitrogen molecules enter the oxygen chamber, and vice versa. After a few minutes they are thoroughly mixed. The molecules keep on bouncing around – just like the billiard balls – but they will never again take on their original configuration. The first chamber will never again be pure nitrogen, the second chamber never again pure oxygen. If the molecules were somehow colored – or if we were talking about milk and coffee instead of nitrogen and oxygen – the film would plainly show that mixing. And again the film makes sense in only one direction: the two components would never "unmix."

The principle at work in these examples has come to be known as the *second law of thermodynamics*. The second law says that the amount of disorder in a closed system – the amount of entropy – can never decrease.* It must always increase, or at best remain the same, over time.

We should take particular note of two aspects of the second law. First of all, it is statistical in nature; it doesn't apply to lone billiard balls, but it does let us characterize the motion of an array of many such balls. Secondly, it reveals a kind of built-in asymmetry to processes that involve large numbers of particles or bodies. That asymmetry is often thought of as indicating the flow of time. In 1927, the British astronomer Arthur Eddington (1882–1944) used the phrase "the arrow of time" to describe this asymmetry, and the vivid metaphor is still with us. (For Eddington, the second law held "the supreme position among the laws of nature.")

The statistical nature of the second law helps us understand why complex systems evolve in one direction only. There are billions of ways to arrange ten billiard balls randomly on a table – but there are by comparison only a few ways to arrange them in a compact triangle (the arrangement they had before the break). As the balls move randomly

---

* A "closed system" is one which is isolated from outside influences. The billiards table, in fact, is *not* such a system; after all, the players keep striking the cue ball, and they cause a *decrease* in entropy, first by racking up the balls and later by sinking them. To better picture the second law of thermodynamics in action, we could imagine isolating the table immediately after the cue ball is struck for the first time. If the table were frictionless and had no pockets, entropy would gradually increase over time, until the balls were randomly scattered across the table.

The Second Law of Thermodynamics

a) A barrier separates two compartments. The first (**A**) contains molecules of nitrogen (black), the other (**B**) contains oxygen (white).

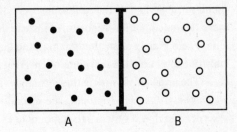

b) With the barrier removed, some nitrogen molecules are now found in **B**, and some oxygen molecules in **A**.

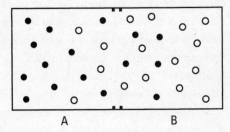

c) Soon the two kinds of molecules are thoroughly mixed. The sequence is irreversible: the two gases will never "unmix." Such processes, guided by the second law of thermodynamics, seem to be linked to the "arrow of time."

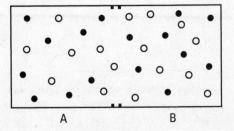

over the course of time, we can see why it is staggeringly unlikely for the balls to take on that original arrangement. The same argument can be applied to the molecules of gas in the container. Or, to take another example, to a shattered teacup. There are many ways for a teacup to smash, but only one way to piece it back together again. A shattered teacup *could*, in theory, spontaneously reassemble – it is not forbidden by Newton's laws – but it is astronomically improbable, thanks to the second law of thermodynamics. In practice, it is so unlikely that we would harbor no expectation of ever witnessing such an event. In all of these examples – and in any complex system – the arrow of time reveals itself as a change in the amount of disorder in a system, a change that always proceeds from lower to higher entropy.

The second law of thermodynamics can feel a little depressing. It suggests, perhaps, that "order" is fleeting – that no matter how often we clean the basement, for example, entropy will have the final victory. Our bodies will inevitably yield to it; perhaps, in the end, so too will our civilization. The Oxford chemist Peter Atkins captured this pessimism when he wrote:

> We have looked through the window on to the world provided by
> the Second Law, and have seen the purposelessness of nature ... All
> change, and time's arrow, point in the direction of corruption. The
> experience of time is the gearing of the electrochemical processes
> in our brains to the purposeless drift into chaos as we sink into
> equilibrium and the grave.

———

The second law of thermodynamics certainly seems to be telling us something about the arrow of time – but perhaps the connection isn't as straightforward as it appears. Have we really explained the distinction between past and future? Consider again a low-entropy situation such as the billiard balls occupying a tight configuration on one side of the table. It is certainly true that, projecting forward in time, we expect to see a higher-entropy situation – a more disordered array of balls. But remember, the laws that we're using to predict the motions of the balls are perfectly symmetric. If we start with the same configuration and try to guess what the most likely arrangement was in the past – if we try to "predict" how a film-played-backward of the last few minutes would appear – then once again we would predict higher entropy. In other words, starting from some particular low-entropy state, the statistical argument that we've been using tells us that entropy increases not only into the future, but into the *past* as well. This is a subtle but profoundly important point: the same analysis that we used to predict that a low-entropy system will evolve into a high-entropy system in the future could be used to "predict" that a low-entropy system *was probably preceded by a high-entropy system.*[*] But we would almost always be wrong. Most tight arrays of billiard balls were in fact preceded by somebody racking them up, not by balls randomly rolling into that arrangement; most unbroken teacups were preceded by someone pouring tea into a cup, not by scattered shards of china spontaneously coming together. Our analysis is clearly incomplete.

The second law of thermodynamics says that if you currently have a low-entropy system, you can expect to have a high-entropy system in the future. But it doesn't say *why* you might expect to have a low-entropy system right now. Perhaps we need to look further back in time. What events have transpired in the past that would lead to the low-entropy universe we find ourselves in today? Perhaps the roots of the second law of thermodynamics can only be understood in the broader context of cosmology – that is, in terms of the origin and evolution of the universe. Maybe the universe started off in a state of very low entropy, and that explains why our observations today show entropy on the rise. This idea comes with its own difficulties, and we will return to it in Chapter 11. At the very least, though, we have learned to be cautious in stating that the second law of thermodynamics "explains" the arrow of time.

---

[*] This is sometimes called "Loschmidt's paradox," after Austrian scientist Josef Loschmidt (1821–95).

Philosophical Interlude, Part One

Philosophers, not surprisingly, have been wrestling with time's many puzzles even longer than physicists have. (Of course, in the era of Newton and Leibniz, "philosophy" encompassed just about everything; what we now call science was embraced by the study of "natural philosophy.") Many of the main themes have endured over the centuries, and some of today's arguments can be traced back to ancient times. The Greeks tied themselves into knots over the problem of motion: How can an object change and still be the same object? Is "time" just another word for "change"? For many Greek thinkers, time was indeed a secondary concept to the more central ideas of change and motion. Heraclitus – we met him in Chapter 4 – was one such thinker. For Heraclitus, change was paramount; everything was seen as being in flux. He famously proclaimed that "a man can't step into the same river twice": it is not the same man and it is not the same river. Parmenides disagreed. In fact, he came to the remarkable conclusion that time doesn't exist. Parmenides' argument goes like this: Anything that we can conceive of must have a permanent existence. Anything that we can speak of, or think of, "is without creation or destruction; whole, unique, unmoved and perfect." For Parmenides, this line of reasoning ruled out change: whatever is, simply *is*; change must be an illusion.

The idea that much of what we perceive is illusion is echoed in the work of Plato. However, Plato did not simply dismiss our flawed perceptions but struggled to see how they relate to the deeper truth of the world that surrounds us. For Plato, time could not be rejected as an illusion, and, with great effort, he integrated time into his cosmology – his picture of the universe – which we looked at briefly in Chapter 4.

Aristotle, as we've seen, believed that time requires change – a view that would surface again with Leibniz two thousand years later. Aristotle also struggled with the concept of "now." He saw the passage of time as a succession of "nows" – a succession of instants – and yet he did not see time as *composed* of such instants. He believed time could no more be composed of "nows" than a line could be composed of points. (No two points can ever touch, he reasoned, because no matter how close they are, we can always insert another point in between.) But Aristotle seems to stumble on the question of what these "nows" actually do. Are they static? Is every instant just like every other instant? Or does "now" somehow flow through, or with, time? Once again the question of time's apparent flow defeats even the most careful thinkers.

Is the Future "Real"?

Aristotle also wrestled with the distinction between past and future. Is the future "real"? It certainly *seems* less real than the present or the past. The future appears indeterminate; future events cannot be spoken of with the same certainty as present or past events. The future is like a mirage: while the past seems etched in stone, and the present is right in front of us, the future looms like a fog of infinite possibility. Aristotle thought very carefully about statements such as "There will be a sea-battle tomorrow." It seemed that the truth of such a statement was undefined; it was contingent on whether the generals would indeed choose to do battle the next day. He reasoned that statements about future events are neither true nor false. One proposition may be deemed more likely than another, but that is as definite as we can be. The future, for Aristotle, has only *potential* existence.

Christian theology embraced many aspects of Aristotle's world view, but also adapted them to its own needs. Augustine, in his *Confessions*, tells of his struggles with philosophy in parallel with his spiritual quest, one that would lead to his adoption of Christianity. Augustine considers the Judeo-Christian view of the creation of the world – and then immediately realizes that it raises difficult questions about time and eternity. Did God create the world "in time"? If so, what was God doing beforehand? As we've seen, this is the same dilemma that would grip Leibniz more than thirteen centuries later, and Augustine reaches the same conclusion: God, he reasons, must have created time *along with* the universe.

Augustine goes on to ponder the distinction between past, present, and future, and puzzles over the apparent "flow" of time. In the end he appears to grant "now" a special status: it is more real, somehow, than the past or future. Indeed, he argues that past and future are, in a sense, embedded in "now," for the present contains our memories of the past as well as our expectations of the future. The extreme version of this stance – the rejection of past and future – is sometimes called *presentism*. Whatever we make of Augustine's line of reasoning, his linking of time to human consciousness is a profound notion; we have already encountered it briefly with Descartes, and it will recur often in our story.

Two Kinds of Time

Anyone who has thought carefully about time will have noticed that we speak of time in two distinct ways. When we talk about what has already happened, what's happening now, and what may yet happen, we build up a mental picture of these events relative to the moment we're currently experiencing. We introduce the terms "past" and "future" to describe how those events relate to the present moment, and the verbs that we use to describe some particular action take on the appropriate "tense" as needed: "I cooked some pasta"; "I'm eating dinner"; "I will wash the dishes." These statements are all relative; the phrase "I'm eating dinner" stops being true once I finish eating. It describes events relative to "now," and "now" keeps shifting.

But when we imagine time as a line – as Galileo and Newton did – then we use a different approach. We attach labels to various events to indicate where they occur on that timeline. It is a much more static view, and we use language differently when we describe events from such a perspective. (Actually, even using the word "static" is a little misleading; it suggests that some kind of meta-time lurks in the background. Instead we must imagine a *timeless* array of events.) When we say "The Declaration of Independence was signed on July 4, 1776," we don't need to say that this was in the past relative to the present moment; when we say "There will be an eclipse in 2017," we similarly don't need to say that this is in the future relative to now. Rather, these seem to be unchanging facts about the world.

The British philosopher John McTaggart (1866–1925) spelled out this distinction in his influential essay "The Unreality of Time" in 1908. He called the two ways of thinking about time the "A series" and the "B series." The A series is simply the everyday notion of time in terms of past, present, and future; it is sometimes called the "tensed" view of time. Any event can be located in time with respect to the A series when the speaker states how long ago the event happened – or how long we must wait until it happens. The B series, in contrast, refers to fixed labels that we attach to specific moments in time – 5:00 p.m. GMT on March 30, 2010, for example. (This is sometimes called the "tenseless" view of time.) Events described in terms of the B series can be marked "earlier than" or "later than" each other – but "now" never enters into it.

When we describe events in terms of the A series, our statements seem to be contingent: the phrase "I had eggs for breakfast yesterday" is true if I say it the day after having the eggs, but it may or may not be true if I say it on some other day. Statements described in terms of the B series have a different feel: sentences like "The Declaration of Independence was signed more than a year after the battles of Lexington and Concord" or "The eclipse of 2017 will happen five years after the London Olympics of 2012" seem to have a more permanent truth; they appear to describe unchanging attributes of world history. I have used tenses in the usual way in those sentences, but one could, with a little practice, develop the habit of omitting past and future tenses when using the B series perspective, and just making do with the present tense. One could say, for example: "The Declaration of Independence is signed more than a year after the battles of Lexington and Concord." One can also capture the essence of past and future – absorbing the A series into the B series, so to speak – by elimi-nating any references to "now": instead of "The eclipse will happen nine years from now," say, "The eclipse occurs nine years after this utterance." It is not our natural way of speaking, to be sure, but philosophers would argue that it conveys the same information.

The world imagined according to the B series can seem quite strange at first. The B series grants no special status to "now"; every point along the timeline is on an equal footing. The word "now" becomes like the word "here" – it means something relative to the person saying it, but it has no absolute meaning. When events are strung out according to the B series picture, we end up with something like a block of time; indeed, the picture is often referred to as the "block universe." It is a short step from adopting the block view of time to the conclusion that the flow of time – and perhaps even free will – are illusions. In the B series view, one could argue that future events – the accession of the next pope, for example – are already "fixed" in time. Are future happenings then simply inevitable? Are they laid out in front of us, with no room for choice? McTaggart, in fact, went even further. As the title of his essay suggests, he eventually concluded – as Parmenides had – that time itself is an illusion. He reasons as follows: logically, past, present, and future are incompatible properties – an event can only be "in" one of them. And yet every event *does* seem to take on all three: the death of Queen Anne (to use McTaggart's example) was once future, then present, and is now past. And so he rejects the "tensed" view of time. But since he believes the tensed view is the only way to address time's apparent flow, he concludes that time is not a meaningful entity.

McTaggart's argument has been debated endlessly in the century since its publication. Many philosophers, even if they don't share McTaggart's ultimate conclusion, have been deeply influenced by the "block time" picture suggested by the B series. In fact, while the A series description seems closer to our intuition, the "timeless" view embodied in the B series probably represents the majority view among physicists and philosophers today – especially in light of Einstein's new picture of space and time, which we will come to shortly.

The Sage of South Newington

One of the strongest supporters of the "timeless" picture is English physicist Julian Barbour. Now in his mid-seventies, Barbour earned his Ph.D. on the foundations of Einstein's theory of gravity, known as general relativity, from the University of Cologne in 1968. Since then he has worked as a freelance theoretical physicist, with no academic affiliation; he supports his family in part by translating Russian scientific journals. In his book *The End of Time: The Next Revolution in Physics* (1999), Barbour argues that time – along with motion and change – is nothing more than an illusion.

There are echoes of Parmenides and McTaggart in his argument, but Barbour has an enormous advantage over his predecessors: he has a deep understanding of modern physics, having worked extensively on both general relativity and quantum theory (both of which we will examine more closely in the next chapter).

I met with Barbour at his home, a 350-year-old thatched-roof farm-house in the bucolic village of South Newington, in the north of Oxfordshire. He invites me into the garden, where we sip tea and begin to talk about space and time, motion and change, Mach and Minkowski. He strikes me as part savant, part English country gentleman.

It is late spring, the sun is shining brightly by British standards, and the scent of wisteria fills the air. Next door is a Norman church dating from 1150; as one of the caretakers, Barbour has his own key. He often escorts visitors through the chapel as they take in its many treasures, including a graphic fourteenth-century depiction of the murder of

Orion Books

Independent physicist Julian Barbour.

Thomas Becket by agents of King Henry II in 1170. (More visitors come to see the historic church than to discuss time, Barbour assures me.) When one too many airplanes has flown overhead – just about the only reminder that we are in the England of the second Elizabeth rather than the first – we move inside.

We pause briefly for a dinner of asparagus, bread and butter, cheese, and fresh strawberries. Then we relax on Barbour's living-room sofa. He leans back, and speaks in a careful, measured voice – a mix of solemn scholarly wisdom and disarming English charm.

Part of the problem with "time," he explains, is that our two best theories – general relativity and quantum theory – treat it very differently. "It's like two children sort of quarreling over a toy they want," he says. "But the trouble is, each wants something different." He believes the only solution is to remove the toy. We have to abandon the notion of time.

The heart of his argument is easy enough to follow. Barbour imagines every "now" as a complete, self-contained universe – a universe in which time is notably absent. If we imagine the history of the universe as frames in a movie, each frame must be considered equally "real." (He refers to the total collection of all "nows" as Platonia, in honor of Plato and his vision of eternal, changeless forms.)

And what of the past and the future? For Barbour, they are every bit as slippery – and unnecessary – as time itself. We have no actual evidence of the past other than our memory of it (and the various artifacts that record those memories, whether natural, like fossils, or artificial, like newspapers). Similarly, we have no evidence of the future other than our belief in it. It is all an illusion.

"From the physicist's point of view, there isn't any sort of 'flowing' of time, and there isn't any sort of 'now' creeping up through the world," Barbour says. "The idea of a 'flowing time' is some sort of illusion that's created, somehow, by our consciousness."

So why do we imagine time as a steady progression, with a relentless "moving now"? Presumably because of the way our minds – and in particular our memories – work. What newspapers and fossils do for history at large, Barbour explains, the human brain does for personal memory. Each of these can be thought of as "time capsules" – physical systems arranged in a very precise way, enabling them to "preserve" the past within them. He calls this "the single most remarkable fact about the world as we experience it." In a certain sense, Barbour is obviously at least partially right: if neuroscience progresses to the point where the brain's neural activity can be "read" in precise detail, one could deduce a friend's life history by studying his brain as it is now; the past would be present, so to speak.

"And, if you think about it, this is exactly what modern geology and modern cosmology are doing at the moment," Barbour adds. Cosmologists infer the universe's past by examining the sky as it is now; geologists do the same for the earth's past. It's all essentially "one snapshot," he says.

Among the ramifications of treating every "now" as equally real is a certain kind of immortality. It's not the sort of life after death that most of us would probably prefer; instead, it's something like life alongside death. Since time does not pass, we do not age. "The instant can't age – the instant is what it is," Barbour says. "So to this extent, there is a Julian at seventy – which I now am – but there's a Julian at sixty, who is just as real as the Julian at seventy. You don't have any sense in which yesterday is any less real than today."

What, then, of time?

"It's a mistake the mind has made," he says. "I would say of time what Laplace said to Napoleon about God: 'I have no need of that hypothesis.'"

On the train ride back to Oxford, the conductor calls out the stations. The platforms all have digital clocks, and the stops (at least in theory) are scheduled to the minute. Time certainly *seems* real. I'm as confused about time as I was before (there I go again, using words like "before"). It's hard to imagine time not being real when we can barely construct a sentence without invoking it in one way or another; the illusion of time – if that's what it is – seems to shape our every thought, and certainly permeates our language. I'm not alone in my confusion. The philosopher Simon Saunders, reviewing Barbour's book for the *New York Times*, called it "gold," adding that, as pedagogy and analysis, it is "a masterpiece" – but he also issued "a philosophical health warning," admitting that he was not sure "if it really makes sense."

Perhaps some future (whoops!) writer will find a clearer way of illuminating the illusion of time. After all, those who came after Newton were often much better at explaining his mechanics than he was. (Undergraduate physics students use a variety of textbooks today; the only one they would *never* use is the one Newton wrote.) Besides, Newton – like all of us – was partly a product of his time. A twenty-first-century presentation of his physics would involve, first of all, stripping away the theological overtones. (Through his idea of absolute time and space, Barbour told me, "Newton had thought that he had seen the anatomy of God. I think he felt that he had almost made the invisible God visible.") Perhaps we, too, are so heavily burdened with preconceived notions of time that we cannot readily shed our biases.

In the meantime, I vow to reread *The End of Time*. It could be worse – at least I'm not trying to plow my way through the *Principia*. I'm reminded of a story – probably apocryphal – about a student who is said to have watched Newton pass by in his carriage. He supposedly quipped, "There goes the man that writ a book that neither he nor anybody else understands."

ALBERT'S TIME

Spacetime, relativity, and quantum theory

Relativity has taught us to be wary of time.

– physicist Wolfgang Rindler, who coined the term "event horizon"

I see the Past, Present, and Future existing all at
once before me.

– William Blake

The city of Bern, nestled between the mountains of western Switzerland,
looks much as it did a century ago. Streetcars still wind their way up and
down the main streets, while arcades line the narrow lanes of the
Aldstadt, or old city. The roads are dotted every few blocks with colorful
fountains, many of them dating to the sixteenth century. Visitors who climb
the gothic spire of the city's cathedral – one of the tallest in the country –
are rewarded with sweeping views of red-tiled roofs, church steeples, and
the churning blue waters of the Aare River. Apart from the cars and the
tourists, in fact, the Swiss capital has changed little since the winter of 1902,
when Albert Einstein, aged twenty-two, arrived here – unemployed, traveling on foot, and carrying all his belongings in a single suitcase. Within
three years, he would become a husband and a father, and – oh, yes –
would develop a radical new picture of time and space that would change
the world forever.

Albert Einstein (1879–1955) wasn't born here – that honor goes to the southern German city of Ulm, some 250 kilometers away. Nor did he remain here for very long; in less than a decade, with his genius bringing increased fame, academic positions would draw him to Zurich, Prague, and Berlin, before the rise of Nazism would force him to leave Europe for good. But it was in Bern that the ambitious young scientist got his first real job – an entry-level position examining patent applications in a government office. And it was here that Einstein had his first great insights into the nature of the universe.

A narrow wooden staircase leads up from the street to the small apartment where he once lived, at Kramgasse 49. It is now a museum, and tour guide Ruth Aegler meets me at the top of the stairs. In Einstein's day, she explains in a thick Swiss-German accent, the apartment consisted of just two rooms, though one of them had two large windows looking out onto Kramgasse. (The museum is somewhat larger, having absorbed several adjacent rooms as well as the floor above.) If Einstein had poked his head out of the window and glanced to the left, he would have seen the magnificent sixteenth-century *Zytglogge*, or clock tower, just a block away. Ornate and colorful, if rather squat, it would have greeted young Albert every time he left the apartment.

Visitors now stream through the modest, wallpapered rooms, pondering Einstein's wooden desk from the patent office, dozens of historical photographs, his doctoral thesis, and even his high-school report cards (which show that he wasn't a bad student after all, contrary to popular myth). The patent job brought in just 3,500 francs a year – barely enough to cover the rent and provide the basic necessities of life for the scientist and his wife, Mileva. "Einstein was so happy and so proud that he could, for the first time in his young life, rent an apartment like that," Aegler says. "Only sixty square meters – that's not much, but for him it was absolute luxury."

By day, Einstein pored over the hundreds of patent applications that passed across his desk. His true passion, though, was not gadgetry but the underlying theory – the machinery of the cosmos itself. By the spring of

1905, a new theory of space and time had taken shape in Einstein's mind. At the age of twenty-six – only slightly older than Newton was at the time of *his* great insight – Einstein had his own *annus mirabilis*, producing five groundbreaking physics papers, including the one that gave us the first part of his theory of relativity, known as special relativity.

The Roots of Relativity

At the start of the twentieth century, Newton's laws seemed to explain just about everything – but not *quite* everything. To understand electricity and magnetism, as well as light and radio waves, physicists relied on another very successful description of nature – a framework developed by the Scottish-born physicist James Clerk Maxwell (1831–79).* Maxwell developed a set of equations that describe the relationship between electric and magnetic fields, showing them in fact to be two aspects of the same phenomenon. Just as Newton had linked celestial and terrestrial mechanics, Maxwell showed that electricity and magnetism are intimately connected. Electromagnetism was also found to embrace light, which was now seen to be an electromagnetic wave – an oscillating electric and magnetic field. (Light is just one kind of electromagnetic wave: X-rays, microwaves, and radio waves are all examples of "electromagnetic radiation," differing only in wavelength.)

But Maxwell's equations suggested something remarkable about these electromagnetic waves: they seemed to travel at one particular speed, which physicists denote by the symbol c – the speed of light. (The speed of light was first accurately measured in the 1670s by the Danish astronomer Ole Römer. The modern value for c is about 300,000 kilometers per second.) The notion that light was a wave that traveled at a definite speed raised two rather troubling questions: First of all, *relative*

---

* A more detailed look at Maxwell and an overview of nineteenth-century physics can be found in Chapter 4 of *Universe on a T-Shirt*.

to what do light waves travel at the speed *c*? Wouldn't the speed of light depend on how you measured it? Surely your own speed, as well as the speed of the light-emitting object, would affect the value that you measured. Did Maxwell's equations apply only in some special reference frame associated with waves of light?

Secondly, just *how* did light waves propagate from one place to another? Everything that scientists understood about waves up to that point suggested that they require some sort of medium in order to move. (Sound waves, for example, require air; waves in the ocean require water.) But light waves had to be able to reach the earth from the sun, across seemingly empty space. What medium supported waves of light?

The best guess that anyone in Maxwell's time had was that light waves vibrate in a substance called the "luminiferous ether" (which I'll refer to simply as the "ether"). The ether, believed to permeate all of space, was presumably the substance that allows light waves to propagate. (It was also said to be the medium by which gravity exerts its influence. Newton never made it clear how the gravitational force of one body was felt by another, distant body; his opponents mocked his notion of gravity for its mysterious ability to reach out across empty space.)

The ether, then, was a convenient solution to both problems: it would give light something to propagate through, and it would define the reference frame of Maxwell's electromagnetic waves. But this still seemed awkward. Shouldn't the laws of physics be the same for everyone? If electromagnetism demands a special, privileged frame of reference, it would seem to violate a very basic principle that goes back to the time of Galileo. Often called the "principle of relativity" (or "Galilean relativity"), it states that there is no "privileged" reference frame: one observer is no better able to measure "true" speeds or distances or intervals of time than any other. Galileo, in fact, described a thought experiment that emphasized that very point: imagine, he said, that you and a friend are cooped up in a windowless cabin on board a moving ship. Suppose you have with you some butterflies and birds, an aquarium full of fish, and a bucket that slowly drips water from a hole at the bottom. You also have a ball that you throw back and forth with your friend. When the ship is tied up at the dock, the animals are equally inclined to move in any direction, the dripping water falls straight down, and throwing the ball takes the same effort regardless of who's throwing it to whom. But – this was Galileo's insight – you would observe *exactly the same effects* even if the ship were in motion at a constant speed. "You will discover not the least change in all the effects," Galileo declared, "nor could you tell from any of them whether the ship was moving or standing still."* Indeed, Galileo had shown that terms like "moving" and "standing still" are merely labels; no observer is in a more privileged position to say that he is at rest (or moving at some particular speed) than any other.

But in Maxwell's electromagnetism, there *does* seem to be a privileged reference frame – the reference frame of the mysterious ether. It would help, of course, if the properties of the ether could be measured in some way – or even detected, for that matter. Physicists had tried to discern the effects of the earth's motion through the ether as our planet revolves around the sun. As of 1905, however, all attempts to detect the ether had failed.

Most scientists were not losing sleep over this apparent dilemma. In the 1880s, Heinrich Hertz had used Maxwell's equations to predict the existence of radio waves, which he then duly detected; in less than a decade, Guglielmo Marconi was busy building radio transmitters and receivers. But a few physicists, including the young Einstein, were troubled by what they saw as a fundamental flaw in the underlying description of nature embodied in these two conflicting world views. (The French mathematician Henri Poincaré and the Dutch physicist Hendrik A. Lorentz were also grappling with these difficulties.)

---

* For most of us today, a jetliner at cruising altitude is probably the most vivid example. As long as the pilot avoids any turbulence or sharp turns, a passenger who closes the window shade might hardly be aware of the plane's motion.

One of Einstein's great strengths was his ability to conjure up "thought experiments" – simple mental pictures that allowed him to visualize what otherwise might seem like quite abstract problems. Even as a teenager, he had puzzled over what sounds like a very simple question: What would happen if you could catch up to a beam of light?* Newton and Maxwell offer starkly different answers to this question. In the Newtonian framework, you can catch up to anything; just go faster and faster, no problem. But in Maxwell's picture, light always propagates at 300,000 kilometers per second. If you caught up to a beam of light, its speed (relative to you) would be reduced to zero. Would you then see "frozen" waves of light? (When a surfer rides the crest of a wave in the ocean, he and the wave travel at the same speed; he could perhaps describe his wave as frozen. But what does frozen light look like?) If you *could* observe frozen light, you would have an "absolute" indication of your speed – a clear violation of Galileo's principle of relativity. To Einstein the thought of a stationary beam of light seemed crazy. "There seems to be no such thing," he said, "neither on the basis of experience nor according to Maxwell's equations." Einstein balked at the notion that an observer moving alongside a beam of light might need a different set of equations – different laws – to describe what he saw. After all, motion is relative; echoing the words of Galileo, Einstein asked how the observer would "know, or be able to establish, that he is in a state of fast uniform motion?" And if frozen light had no meaning, what *would* happen as your speed increased, approaching the speed of light?

Historians continue to debate exactly how Einstein arrived at his solution. The job at the patent office – as menial as it seems to us in hindsight – probably gave him valuable mental exercise as he imagined which electrical gadgets would work and which would falter. Historian Peter Galison has argued that the problem of synchronizing Europe's

---

* Einstein first pondered this "thought experiment" at the age of sixteen, but it is clear that it continued to influence his thinking in the years leading up to his 1905 paper on special relativity.

electrical clocks was especially important; many of the patents that crossed Einstein's desk involved electronic devices linked to this problem. Einstein's conversations with a close group of friends in Bern, nicknamed the Olympic Academy, gave him an invaluable sounding board for his ideas about space and time; his wife, Mileva Marič – a former classmate – also played such a role. (In a typical love letter from a few years earlier, Einstein tells Mileva how he looks forward to the hikes they'll take together when he comes to visit her in Zurich: "The first thing we'll do is climb the Ütliberg," he writes, referring to a local hill. "I can already imagine the fun we'll have . . . and then we'll start in on Helmholtz's electromagnetic theory of light.") Einstein was also profoundly influenced by the philosophical works of such thinkers as David Hume and Ernst Mach, whom he read intently in what little free time he had.

Whatever triggered it, the answer came to Einstein in the first few months of 1905 – seemingly out of the blue, but in fact the culmination of a ten-year period of intense mental focus in which he seems to have thought of little else. The solution, as he told his friend Michele Besso that May, "was to analyze the concept of time." Time, he goes on, "cannot be absolutely defined, and there is an inseparable relation between time and signal velocity."

Einstein had found that Newton's laws had their limitations; they were only an approximation of the true picture. Newton's equations apply correctly as long as the velocities involved are low: at everyday speeds, they're perfectly adequate. But as one approaches the speed of light, they break down. A new kind of framework is needed.

Einstein's paper, titled "On the Electrodynamics of Moving Bodies," was received by the prestigious journal *Annalen der Physik* on June 30, 1905. It was thirty pages long, but Einstein had already overthrown the Newtonian world view within the first few pages, introducing a new way of looking at both time and space, and, for good measure, eliminating the ether once and for all. (The reason no one had detected it, Einstein reasoned, is because it doesn't exist.) At the end of the paper there were no references to earlier work by other scientists, though he thanked his friend Besso "for several valuable suggestions." With his June 30 paper, Einstein had finally shown how to reconcile the mechanics of Newton with the electromagnetic theory of Maxwell.

A New View of Time and Space

The theory that Einstein laid out in 1905 is known as the special theory of relativity, or "special relativity" for short. It rests on two assumptions, or "postulates." The first postulate says that the laws of physics must be the same for any two observers, no matter how fast they're moving relative to one another, so long as they're moving at constant speeds (that is, with no acceleration). Whether you're trying to predict the motion of a projectile, or measuring electrical or magnetic properties, or studying a beam of light, the laws must be the same for everyone.

The second postulate says that the speed of light is always the same, regardless of your own speed and the speed of the object that's emitting the light. In other words, you'll always measure a beam of light as traveling at a specific speed, c.

The first postulate isn't particularly radical. It basically restates Galileo's principle of relativity – the notion that there is no privileged reference frame – and raises it to the status of a fundamental postulate about the physical world.

The second postulate, however, is truly shocking. In Newton's world, the speed you measure for any object depends on its motion and on your motion. A train seems to be whizzing past if you're standing on the platform; if you're on board the train, however, it doesn't appear to be moving at all (while the platform appears to be zooming past in the opposite direction). Throw a baseball off the front of the train, and the observer on the platform sees the ball as being given a "boost": if the train is going 100 km/h and you throw the ball at 80 km/h, an observer on the ground will see the ball moving at 180 km/h. The math couldn't be simpler: the speed measured by the observer on the ground is the sum of the train's speed and the ball's speed. It's just $v = v_1 + v_2$. Sounds obvious, right?

In Einstein's theory, this is still *approximately* true of trains and base-balls and other slow-moving objects – that is, objects moving at much less than the speed of light. But his second postulate says it is *not* the case for light: no matter how fast you're moving, and no matter how fast a source of light is moving, you'll still measure that beam of light as traveling at 300,000 km/s. It doesn't matter if I'm walking slowly with a flashlight or if I've mounted the flashlight on a rocketship, speeding past you at, say, 200,000 km/s (two-thirds the speed of light). It makes no difference; you will still measure that beam of light as moving at 300,000 km/s. This is the final answer to Einstein's thought experiment about catching up to the speed of light: it can't be done. Light can neither be given a boost nor slowed down; and no matter how fast *you* move, the beam of light will still appear to be traveling at its usual speed. (This also has the effect of making the speed of light the ultimate "speed limit" in the universe.)

In fact, whenever speeds are comparable to the speed of light, they no longer add up as simply as they did in Newton's world. The speed v that you measure is no longer equal to $v_1 + v_2$. Einstein worked out the correct formula.* It still gives Newton's result at low speeds, but at high speeds you get a lower sum than you would have expected. (And no matter how large v_1 and v_2 are, the sum will never be greater than c.)

And now the showstopper: in order for the speed of light to be constant, *time and space must be relative*. In other words, if two observers are moving relative to one another, they can disagree about the time interval between two events or the distance between two points in space – something that could never happen in Newton's arena of absolute time and space. And Einstein showed how to calculate those discrepancies precisely.

How can two observers, with perfectly synchronized clocks, possibly disagree about the interval of time elapsed between one event and

---

* The new formula is $v = \dfrac{v_1 + v_2}{1 + \dfrac{v_1 \times v_2}{c^2}}$

another? This is perhaps the most counterintuitive implication of Einstein's theory. In fact, it turns out that a clock moving at high speed will appear to "tick" more slowly than an identical clock that's "stationary" (the quotation marks here are just a reminder that either of the two clocks can be said to be the one that's moving). This effect is known as *time dilation*. As an example, consider an imaginary high-speed train carrying a "light beam clock" – a clock made from a pair of parallel horizontal mirrors, one above the other, with a beam of light bouncing up and down between them (page 162, upper diagram). When the train is stationary, a person on board and a person standing on the platform both measure the same interval of time between each "tick" of the clock. When the train moves at close to the speed of light, however, an observer on the platform sees the beam of light trace out a diagonal or sawtooth path (lower diagram). Therefore, the distance the beam of light travels during each "tick" is larger. But – and this is the crucial part – Einstein's second postulate requires that the speed of light is still measured as having the same value. Since speed is equal to distance divided by time, and since the distance is larger, the time between each "tick" must increase. Therefore, an observer on the ground sees the clock on board the moving train as running slow.

But the observer on board the train comes to the opposite conclusion: he sees a light beam in a similar clock on the ground as tracing out a diagonal path, and concludes that *it* is running slow. In keeping with Einstein's first postulate, there is no special, "preferred" reference frame – both descriptions are equally valid.

The time-dilation effects are negligible at everyday speeds but become significant as one approaches the speed of light. (I won't bother with the formula for time dilation here, but it involves only junior-high mathematics – some subtraction, division, and a square root.) If your friend's rocket is whizzing past at eight-tenths of the speed of light, you will observe her clock to be ticking at only 60 per cent of its usual rate. At nine-tenths the speed of light, it slows to just 43 per cent of its normal rate; at 99 per cent of light speed, it falls to 14 per cent. (Your friend

Why Time Is Relative

a) "Light beam" clock on board stationary train

Clock measures time by means of light pulse
moving up and down between two mirrors.

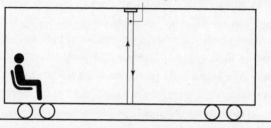

The passenger in the train and an observer on the ground
measure the same interval during each cycle of the clock.

b) Train moving near speed of light

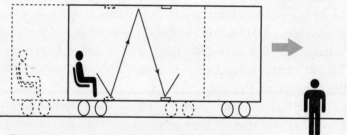

The observer on the ground now sees the light pulse trace
out a diagonal path. Since the speed of the light pulse is
constant, he measures a greater interval during each cycle of
the clock — and concludes that the clock is "running slow."

• He also sees the train (and everything in it) become shorter
— but only in the direction of the motion.

cannot reach the speed of light; if she could, you would see her clock as having stopped altogether.)*

Interestingly, the equations that Einstein used in his June 30 paper were not new. They had been known to both Poincaré and Lorentz – but neither man had made the crucial leap of interpretation; neither had seen how the principle of relativity and Maxwell's electromagnetism could be reconciled by thinking about time and space in a new way. Einstein, it seems, had a unique ability to step back and see the big picture.

Perhaps his relative isolation at the patent office – he had, up to that time, failed to secure a job in academia – worked to his advantage; as someone cut off from the physics establishment, he had no particular loyalty to its preconceptions. In other words, he had nothing to lose. "He comes in entirely as an outsider," says Harvard historian Gerald Holton, a leading Einstein scholar. "He has no stakes at all in any of the nineteenth- and the early-twentieth-century physics . . . He lets his mind wander. He's not endangering his academic position, because he doesn't have one, and he can take those risks . . . He takes a much more Olympian view than any of the others did."

The physics community (and eventually the world at large) came to see special relativity as revolutionary, but Einstein never saw it that way. His aim was merely to extend Maxwell's theory of electromagnetism to a wider array of phenomena. Stressing the fact that the speed of light was the same for everyone, rather than the fact that time and space were not, he at first called his idea *Invariantentheorie* – "invariance theory." But Poincaré and the great German physicist Max Planck called it the theory of relativity, and that's the name that stuck.

---

* Two other effects are worth mentioning: the length of a fast-moving object will appear to shrink, and its mass – its resistance to further acceleration – will increase.

The Problem of "Now"

The slowing down of fast-moving clocks is just one of the ways in which special relativity assaults our commonsense view of time. It also forces us to re-think the idea of *simultaneity*. We say that two events are simultaneous if they happen at the same time; in Newton's world, this was a very straightforward idea. But in Einstein's universe, we have a problem: two events that may be simultaneous for me may not be simultaneous for you, depending on our relative motion. This is called the "relativity of simultaneity."

Suppose again that we have a railway car, this time equipped with a couple of simple gadgets: at either end of the car we install a camera's flash unit along with a photodiode, and we wire it up so that when light, from any source, hits the photodiode, it triggers the flash (page 165, upper diagram). (We can imagine that the car is dark enough that unless we introduce extra light, the flash units just sit there and do not fire.) Let's call the unit on the left A and the one on the right B. Now I take up a position in the middle of the car, halfway between A and B. I have a third flash unit, which I'm holding in my hand. If I set off the flash, what happens? Light from the flash reaches A and B at the same time, and forces them both to fire. From my perspective, the flashes triggered at A and B are simultaneous.

Now imagine that the railway car is moving along from left to right, at some speed close to the speed of light (lower diagram). Standing in the middle of the car, I set off my flash and again see the flashes at A and B as simultaneous. (Einstein's first postulate demands this; it simply says that I can equally well describe myself as being at rest, with the station and the platform speeding by.) But what does an observer on the ground see? From her perspective, the back of the car (A) is "chasing" the beam of light, while the front of the car (B) is running away from it. From her point of view, the beam travels a shorter distance to reach A than it does to reach B. According to Einstein's second postulate, she sees the beam itself moving at the usual speed, c – and concludes that the time needed for the beam to reach A, and trigger its flash, is less than the time needed for the beam to reach B. In other words, she sees the flash at A before she sees the flash at B. The events are no longer simultaneous.

Why Simultaneity Is Relative

a) Flash emitted from center of stationary train car

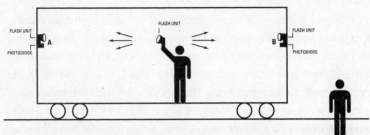

To the passenger and to an observer on the ground, the light
from the flash emitted at the center of the car reaches the
two ends of the car at the same time. The photodiodes
trigger the secondary flash units at **A** and **B** simultaneously.

b) Flash emited from center of car as train moves near speed of light

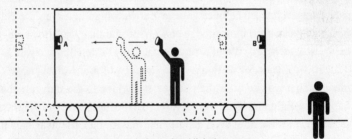

To the observer on the ground, the emitted light beam now
reaches **A** first, as the back of the train "catches up" to the
beam of light. He sees the secondary flash unit at **A** fire
before the unit at **B** fires. The events at **A** and **B** – still
simultaneous for the passenger – are no longer simultaneous
for the observer on the platform.

With special relativity, we can no longer declare two events to be simultaneous in any absolute sense. Instead, we can only say that they may *appear* simultaneous from some particular frame of reference. (Physicist Brian Greene calls this "one of the deepest insights into the nature of reality ever discovered.")

It gets worse. What do we mean when we say a particular event is happening "now"? When we use the word "now," we are really comparing two events: I can snap my fingers and then ask whether some other event is simultaneous with my finger-snapping or not. If it is, I say that the event is happening "now." In the Newtonian universe, I can legitimately ask, "What events in the universe are happening right now?" The answer would be a unique set of occurrences, scattered throughout space but lying on a single "slice through time." I can snap my fingers at, say, noon Eastern Standard Time on December 1, 2009, and every event, everywhere in the universe, either is simultaneous with my finger-snapping or is not. That was fine for Newton, but not for Einstein. As we have seen, in special relativity there is no universal agreement among observers as to whether two events actually are simultaneous or not – and thus there can be no universal "now." As Einstein remarked, "There is no audible tick-tock everywhere in the world that can be considered time." (There is an amusing story about a lecture he once gave in Zurich, where he covered a blackboard with clocks to illustrate the concept of simultaneity. After a lengthy exposition, he asked, "What is the time, actually? I don't have a watch.")

It is difficult for us to abandon the idea of a universal "now." We imagine that we can utter the phrase "everything in the universe that is happening right now" and have it refer to a meaningful set of events. But Einstein shows us that such a statement has, in fact, no clear meaning. Each observer has his own list of events that appear to be happening "now," and no one person's list is more authoritative than the next. There is no "master clock" for the universe that can tell us what happened when. "Now" – one of the simplest and most-often-uttered words in our language – seems to have slipped from our grasp.

Newton would not be pleased, but in the century since Einstein unleashed special relativity on the world, countless tests have confirmed the theory's predictions. A new benchmark was set in the fall of 2007, when a team of scientists led by Gerald Gwinner of the University of Manitoba confirmed the time-dilation effect to one part in 10 million. Gwinner and his colleagues used an accelerator in Germany to whip lithium ions through a circular tube at 6 per cent of the speed of light. They then used a laser to stimulate the ions, forcing them to give off radiation. Because the radiation is an oscillating electromagnetic wave, it can act as a clock; one cycle of the radiation can be thought of as one "tick" of the clock. At the high speeds involved, the ticks slowed down – observed as a lowering of the frequency of the radiation. In Gwinner's experiment, the frequency shift was just that predicted by special relativity.

In the fall of 1905, Einstein published a short follow-up paper that showed another surprising consequence of his postulates – a link between matter and energy, embodied in what is now the most famous equation in the world: $E = mc^2$.*

With special relativity, space and time are much more closely related than Newton could have imagined. In the Newtonian world, two events may be separated either in time or in space, or both. A position in space can be given by three numbers (for example, latitude, longitude, and altitude), while a moment in time can be described by a single reference (for example, by giving the precise time and date). With special relativity, however, we have to imagine combining these two sets of information. We must now think of events laid out in a four-dimensional array that we can call "space-time" – an idea that would be formulated with mathematical precision by Einstein's former math teacher Hermann Minkowski (1864–1909). In a famous proclamation delivered at a lecture in 1908, Minkowski laid the old, traditional view of space and time to rest: "Henceforth space on its

---

* In the equation, E is energy, m is mass, and c once again is the speed of light. Because c is so large – and c^2 is even larger – even a small amount of mass can be converted into a large amount of energy.

own and time on its own will decline into mere shadows, and only a kind of union between the two will preserve its independence."

It is not easy to picture a four-dimensional scene, but if we ignore one of the space dimensions, then it is not too hard to sketch objects from the perspective of Einstein's spacetime. We can plot the remaining two space dimensions along a horizontal plane, and imagine the vertical axis as that of time. The path of a body through spacetime is called its *world line*. A familiar system, such as the earth and the sun, is now seen in a new light (page 169). If we choose a reference frame in which the sun is stationary, then its world line is a straight vertical line, while that of the earth becomes a helix.

There is another useful concept that will aid us in picturing how events in spacetime relate to one another – what physicists call a *light cone* (page 170). Again we imagine space extending along the horizontal plane, and time extending vertically. Suppose a flash of light is emitted at some particular time at a point *P*. (Actually, we should call *P* an "event" – a point isolated in space *and* time.) The rays of light, traveling outward from *P* at the speed of light, trace out a cone-shaped region in spacetime, with *P* at the bottom. This is called the "future light cone of *P*."

Remember, nothing can travel faster than light – so the light cone encloses the entire region of spacetime that a person at *P* could "visit." Indeed, the event *P* cannot have any influence of any kind on the region of spacetime outside the cone. In the same way, we can draw another cone extending downward – that is, backward in time – from *P*. This is *P*'s "past light cone." Only events from within this cone can have had any influence on *P*. Again, this view is in sharp contrast with the Newtonian picture: in the classical view, with no speed limits, you can influence events in any region of spacetime, so long as they lie in the future – and you can *be influenced* by events in any region of spacetime, so long as they lie in the past. Relativity, in contrast, places much of the universe off limits.

In four-dimensional spacetime, what are we to make of notions such as past and future, or before and after? The situation is a bit messier than we might imagine from the clean lines of those light cones. Thanks to the relativity of simultaneity, an event that lies in my past may yet be in your

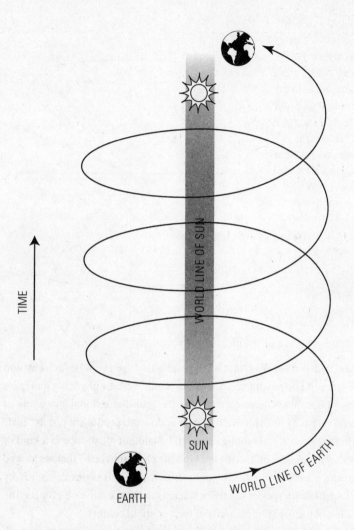

TIME

WORLD LINE OF SUN

SUN

EARTH

WORLD LINE OF EARTH

Picturing spacetime: We can't draw four-dimensional
spacetime – but if we ignore one space dimension, we
can imagine a three-dimensional scene in which time is
the third dimension. In this sketch of the Earth-sun system,
time runs along the vertical axis, while the Earth's "world
line" is seen as a helix.

The "light cone": only events in the "past light cone" of **P** can have had any influence on **P**; and only events in the "future light cone" of **P** can be influenced by **P**. (In this spacetime diagram, only two space dimensions are shown. The vertical axis represents time.)

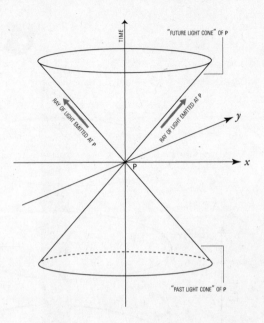

future, and vice versa; all that is required is that the event be far away, and that you and I be moving relative to one another. Many physicists, and quite a few philosophers, have come to see the four-dimensional spacetime of special relativity as evidence that past and future events are just as "real" as present events. Everything seems to be laid out all at once in a kind of block – very similar, in fact, to the kind of "block universe" that we looked at in the previous chapter in connection with the B series described by McTaggart. With special relativity, it seems the block universe now has the backing of the greatest physicist of the twentieth century.

What did Einstein make of this idea? His writings suggest that, much like Parmenides, Augustine, and McTaggart, he viewed the idea of time – or at least the "flow" of time – as something that resides not "out there" in the universe but rather within each of us. "The primitive subjective feeling of time flow," he once said, "enables us to order our impressions, to judge that one event takes place earlier, another later."

Many philosophers seem to agree. The American philosopher Hilary Putnam used the absence of a universal "now" to argue that future events are fully predetermined. Let's say that an event – an election, for example – is in my future, but in your past. (Again, though it may seem odd, this is quite plausible in special relativity, depending on our speeds and the distances involved.) Putnam argues that at the moment when we pass, I am obligated to consider "real" everything that you consider real – including the results of the not-yet-held election! (Not yet held from *my* perspective, that is.) This clearly plays havoc with our traditional ideas of free will and the "openness" of the future. And it's not just the future that gets an overhaul: by a similar line of reasoning, I am obligated to regard as "still real" events that seem to lie in my past but that may be in the present from your perspective.

The philosopher Michael Lockwood appears to be persuaded by Putnam's argument. "To take the spacetime view seriously," he writes, "is indeed to regard everything that ever exists, or even happens, at any time or place, as being just as real as the contents of the here and now." The "block universe" picture, with its peculiar blurring of past, present, and future, continues to hold sway with both philosophers and physicists. To be sure, each point in spacetime has a past and a future, defined by those light cones, but *every* point has such cones – there is no "overall future" and no "overall past." As physicist and writer Paul Davies has put it, "The very division of time into past, present and future seems to be physically meaningless."

Einstein's Masterpiece

Einstein's paper of June 1905 was just the beginning. Special relativity worked only for bodies moving at a constant speed – not those that are accelerating. It also ignored gravity. He struggled in pursuit of a more complete theory, until one day in 1907 he had a breakthrough based on another simple mental picture: "If a man falls freely, he would not feel his weight," Einstein mused. "This simple thought experiment made a deep impression on me."

Einstein was drawn to a profound conclusion about accelerated motion, analogous to what Galileo had deduced about motion at constant speed. Galileo had said that if you're in a windowless vehicle, you have no way of knowing whether you are at rest or moving at a constant speed. Einstein realized that if you're in a similar vehicle and feel a force pulling down on you, you could be accelerating upward *or* you could be experiencing the pull of gravity – the two are exactly equivalent. (Think of how you momentarily feel heavier as an elevator begins to move upward.)

This link between acceleration and gravity turned out to be the key to developing a new framework for describing both phenomena. It took great mathematical prowess, however, to work out the details. Instead of the "flat" geometry of Euclid, Einstein needed the "curved" geometry developed only recently by the German mathematician G.F. Bernhard Riemann (1826–66). "In all my life I have laboured not nearly as hard," Einstein said to a colleague. "I have become imbued with a great respect for mathematics . . . Compared with this problem, the original relativity is child's play."

By the end of 1915, Einstein had worked out an entirely new mathematical description of gravity. This work became known as the general theory of relativity, or "general relativity," to contrast it with the earlier theory of special relativity. The theory depicted gravity in an entirely new way: for Newton, gravity had been a force acting over a distance; for Einstein, it was a warping or curving of space itself. (The most popular analogy is that of a large rubber sheet: When a heavy bowling ball is placed on the sheet, it warps the sheet in proportion to the mass of the ball. Any marbles rolling nearby will be deflected by that curvature.) The sun holds the earth in its orbit by means of just such a warping. In Einstein's world, matter distorts the very fabric of the universe and we experience that distortion as the force of gravity. For Newton, space and time are a kind of static background or stage on which physical events unfold; for Einstein, space and time are themselves dynamic players in the cosmic drama.

General relativity's first success involved the orbit of the planet Mercury. Since the mid-1800s, astronomers had noticed that Mercury didn't trace out a perfect ellipse as it revolved around the sun. Instead, the planet's orbit shifted by a small amount each time it rounded the sun. The effect – astronomers call it *precession* – is very slight, amounting to less than a hundredth of a degree per century. Yet Newtonian physics could not account for it. Einstein's theory, however, correctly described the phenomenon.

Three years later, the theory would be put to an even more crucial test. Because the sun distorts the space around it, it should deflect a ray of light emitted by a star if the ray happens to pass nearby. In other words, when the sun passes in front of a distant star, the star's position – as seen from the earth – should appear to be slightly shifted. The sun is normally too bright to allow such an observation, so astronomers had to wait for a total solar eclipse, in which the moon blocks out the sun's light. Einstein's predictions were confirmed during an eclipse on May 29, 1919; the images of distant stars were indeed displaced, just as the theory predicted. When the results were announced at a meeting in London in November of that year, it made headlines around the world. *The Times* of London declared a "Revolution in Science"; a few days later the *New York Times* buzzed, "Lights All Askew in the Heavens." Einstein would remain a celebrity – albeit a reluctant one – for the rest of his life.

Time, Gravity, and Black Holes

When we think of a heavy object warping the space around it, and imagine the bowling ball on the distorted rubber sheet, we are overlooking something very important: the massive object is warping space *and time* in its vicinity. (Indeed, as general relativity builds on special relativity, we can simply say that it distorts spacetime.) It turns out that, according to general relativity, time slows down in a gravitational field. The stronger the field, the greater the effect (which is known as "gravitational time dilation"). If two hikers with perfectly synchronized watches take a stroll through the hills of Scotland, the one who takes the high road will find his watch running ahead of that of his friend who chose the low road. (In this example, of course, the difference would be far too small for a store-bought watch to register.)

Gravitational time dilation has now been observed in many different kinds of experiments. One of the more dramatic tests was carried out in 1971, when scientists took atomic clocks around the world on board commercial jetliners; they later compared them to identical clocks that had remained on the ground. The experiment is more complex than it seems, because it tests not only general relativity (the clocks on board the planes were higher than those on the ground, and thus in a weaker gravitational field) but also special relativity (the clocks on the plane were moving at a high speed). The scientists were able to disentangle the two effects, and the results were in agreement with Einstein's theory.* Various terrestrial versions of the experiment have also been carried out. For example, an atomic clock in a U.S. lab in Boulder, Colorado – some 1,600 meters above sea level – has been measured to gain about 5 microseconds each year compared with an identical clock in Greenwich, England, which is just 25 meters above sea level. So far, every experimental test of general relativity has confirmed Einstein's predictions. (In fact, GPS devices – Global Positioning Systems – have to take into account the effects of both special and general relativity in order to function.)

In most cases, gravity's warping of spacetime is very subtle, which is why it went unnoticed for so long. (The sun is 300,000 times heavier than the earth; even so, its gravitational field only deflected those light rays during the 1919 eclipse by a meager 1/2,000 of a degree.) But general relativity predicts that more massive objects can warp spacetime much more severely. Black holes are the most extreme example: these exotic structures warp space and time so drastically that they cut themselves off from the rest of the universe.

A black hole can form when a massive star exhausts its nuclear fuel and can no longer support its own weight. It then begins to collapse. If the star is big enough – more than three times the mass of our sun, roughly –

---

* In fact, there was a third effect to disentangle: the earth's rotation causes its own time dilation effect!

then there is nothing that can stop that collapse. Gravity shrinks the dying star, and when it contracts below a certain critical threshold, known as the "event horizon," something peculiar happens: it becomes invisible. A ray of light emitted from within the event horizon can never escape; the gravitational pull is simply too strong. Objects inside the event horizon are effectively cut off from the outside world. (Although black holes cannot be seen, there is very strong indirect evidence that such objects do, in fact, exist. There may even be a "supermassive" black hole at the center of most galaxies, including our Milky Way.)

If an astronaut were falling into a black hole, a distant observer would see her wristwatch slow down almost to the point of stopping. If she were to reach the event horizon, we would see her watch stop altogether. From the astronaut's point of view, time would continue to pass normally as she fell into the hole and, in all likelihood, into a nasty death. (Actually, the intense radiation from the region just outside the black hole, as well as the tidal effects of the gravitational field,* would almost certainly kill her long before she actually crossed the event horizon.) But for us, observing from far outside the black hole, she would remain forever poised on the event horizon, frozen in time.

The entire field of astrophysics has been transformed by general relativity – and so too has cosmology, the study of the origin and evolution of the universe itself. Our understanding of the birth of the universe has been completely overhauled thanks to Einstein's theory – and we will explore this new perspective in more detail in Chapter 9, when we look at the question of how time, and the universe, began.

How should we picture time as described by Einstein's general relativity? There are still world lines and light cones – but the cones can be slanted

---

* "Tidal effects" refers to the difference in the strength of gravity's pull on different regions. If the astronaut were falling into the black hole feet-first, gravity would pull on her feet more than her head, with disastrous consequences.

and the world lines twisted as spacetime itself is distorted by gravity. Even so, many philosophers believe that the framework of the theory is compatible with the block universe we considered earlier in connection with special relativity. (In general relativity, for example, simultaneity continues to be relative.) We should also note that the equations of general relativity are time-symmetric, just like Newton's equations; nothing in relativity directly addresses the question of time's apparent flow.

The Quantum Revolution

It must be a good feeling to know that you're going to win a Nobel Prize. By 1920, now an international superstar and a household name, Einstein knew he was overdue for the prize. (He had in fact been nominated many times, beginning in 1910, but had been passed over for a variety of reasons, some involving politics and some involving science. Many considered relativity too abstract and theoretical – at least before the eclipse of 1919 – to deserve the prize.) He also knew he would not keep the money: he had already promised it to his wife, Mileva Marić, as part of the divorce settlement. (Einstein divorced Mileva in 1919 and married his cousin Elsa Lowenthal later that year.) When he finally won the Nobel in 1921, the citation did not mention relativity – but it did mention one of the other papers he had written during his miracle year in 1905. He was awarded the prize "for his services to theoretical physics, and especially for his discovery of the law of the photoelectric effect."

That paper, written just a few weeks before the one on special relativity, dealt with the way that light interacts with metals, known as the "photoelectric effect." It was a key contribution to a new scientific framework called "quantum theory," which had its roots in a paper by Max Planck from 1900. Planck was struggling to explain the spectrum of the radiation emitted by hot objects.* The mechanics of Newton and the

---

* The *spectrum* of a body refers to the intensity of the radiation given off at different wavelengths.

electromagnetic theory of Maxwell, he found, were inadequate. He was ultimately led to the idea that heat energy is emitted not in a continuous stream but in discrete bundles of a particular (very small) size. Planck called each of these energy bundles a *quantum* (plural *quanta*), from the Latin word for "how much."

We will not wade too deeply into the early development of quantum theory. It will be enough to note that, over the next few decades, quantum theory blossomed into a new system of mechanics that would replace Newtonian mechanics whenever small distances are involved. Newton's mechanics were fine for baseballs and planets (so long as they moved at slow speeds compared to the speed of light), but the atomic and sub-atomic world could only be studied by means of quantum theory. (The old Newtonian picture would come to be known as "classical" mechanics.)

Quantum mechanics is radically different from Newtonian mechanics. It is, first of all, inherently probabilistic. In classical mechanics, a particle is either at position x or it is not. In the quantum picture, we cannot be so precise. We can only say that when we measure the particle's position, we have some particular probability of finding it at x. In fact, quantum mechanics says that, until we measure it, a particle can be in many places at once; more generally, a quantum system can be in many "states" at once – a phenomenon known as *superposition*. It is only when we observe a quantum system – when we make a measurement – that the system "collapses," and we obtain a single, specific result. (At least, that's how the collapse is viewed in the traditional "Copenhagen interpretation" of quantum mechanics. There is another interpretation known as "many worlds," which we will encounter in the next chapter.)

Time and the Quantum

What is the fate of time in the quantum world? One often hears that the probabilistic nature of quantum measurements leads inexorably away from determinism and (allegedly) allows for an "open" future. However, things are not that simple. It turns out that the equation that describes how a quantum state will evolve, known as the Schrödinger equation, is deterministic; it says that quantum systems do, in fact, evolve in a perfectly predictable way. (And the equation is, once again, time-symmetric.) The system evolves in a predictable way *until* we intervene by making a measurement, at which point the wave function collapses. The collapse of the wave function appears to be irreversible, and thus suggests a link to the arrow of time. As Paul Davies puts it, "In the act of measurement, a single, specific reality gets projected out from a vast array of possibilities . . . The possible makes a transition to the actual, the open future to the fixed past – which is precisely what we mean by the flux of time."

While quantum theory may eventually point the way to a deeper understanding of the arrow of time, is also threatens to overturn some cherished ideas about cause and effect. In the classical world, every event has a cause, even if we cannot deduce what that cause is. In the quantum world, some events – the decay of a radioactive atom, for example – can just "happen." This is a radical departure from the idea of cause and effect that we were used to from classical physics. But we may just have to accept it: like relativity, the predictions of quantum theory have been borne out by countless experiments. Every time we use a semiconductor-laden gadget, we bear witness to quantum theory's reach.

"Anyone who is not shocked by quantum theory has not understood it," the great physicist Niels Bohr once said. Einstein himself, in spite of being one of the theory's founders, was never able to fully accept the quantum picture of the world. Had Einstein fully embraced the theory, perhaps he would have been the Person of the Millennium and not just the Person of the Century – but that is idle pondering. In a few weeks in early 1905, he overturned our view of the universe in a single swoop of the imagination, and that is a feat that never ceases to impress. "Newton, forgive me," he wrote many years later, acknowledging that he, too, was standing on the shoulders of giants. "You found the only way that was available at your time to a man of the highest reasoning and creative powers."

In Bern, visitors stroll through the modest apartment where Einstein lived at the height of *his* creative powers – the rooms where he shared simple meals with his young wife, rocked his newborn son to sleep, played the violin, and imagined a new universe. His friends would drop in after dinner, and linger for long conversations. They talked about mechanics and electricity, clocks and reference frames, physics and philosophy. From those swirling discussions, fueled by cigarettes and Turkish coffee, the seeds of relativity took root in Einstein's mind.

Few of today's visitors are professional scientists. Instead, they come from all walks of life, drawn by Einstein's universal appeal not only as a figure of otherworldly intellect but also as a man of enormous compassion, a kind of secular saint for the scientific age. Children sometimes draw sketches in the museum's guest book. Ruth Aegler, the tour guide, mentions a nine-year-old boy who left a note to "Dear Albert," telling of his desire to become "a scientist like you are." Adults can be just as awestruck, she says. Some of them plead for permission to stay past closing time, just to breathe the air that Einstein once breathed. "You can't explain that," Aegler says. "That is something you have to feel."

BACK TO THE FUTURE

The science of time travel

"I know," he said, after a pause, "that all this will be absolutely incredible to you, but to me the one incredible thing is that I am here to-night in this old familiar room, looking into your faces, and telling you all these strange adventures."

– H.G. Wells, *The Time Machine*

When physicist Ronald Mallett was a child, his world revolved around his father, Boyd – a larger-than-life character who worked hard, played hard, and adored his young family. Boyd Mallett had served in the U.S. Army in the Second World War; as a battlefield medic, he was in one of the first units to cross the Rhine in early 1945. After the war, he worked full-time for an electronics company by day and repaired TVs at night and on weekends. His death from a heart attack at the age of thirty-three "completely crushed" young Ronald, who was just ten at the time. The tragedy left him depressed, and his spirits began to rise only when he discovered H.G. Wells's classic novel *The Time Machine* and, a few years later, the theories of Albert Einstein. What he read left him convinced that he might one day see his father again – and made him determined to pursue a career in science.

For years, Mallett, now a professor of physics at the University of Connecticut, dreamt of building a time machine – but he kept his desire under wraps for fear of ridicule. "I wanted to get tenure," he says with a laugh. Fortunately, he found a field of study that would help him understand the nature of time and that was just "legitimate" enough to let him carry out his research without raising any eyebrows. And so he immersed himself in the theory of black holes. "You know, in physics there's this distinction between what people might call 'legitimate crazy' ideas and 'crazy crazy' ideas," Mallett told me recently in his office at the U of C campus in the sleepy New England town of Storrs, Connecticut. "Black holes are what people would consider 'legitimate crazy,' whereas time travel was 'crazy crazy.' It turns out that by studying black holes I could study time."

Mallett remained in the "time-travel closet," as he puts it, until the late 1990s; his seclusion eventually came to an end when the popular magazine *New Scientist* ran a feature story on his work. ("FLASHBACK: Presenting the world's first time machine," the magazine's cover trumpeted.) These days, his website cheerfully lists "time travel" as one of his primary research interests.

Out of the Time-Travel Closet

Black holes, as we saw in the previous chapter, can slow the passage of time by means of their enormous gravitational fields (an effect predicted by Einstein's general theory of relativity). But Mallett realized that light, too, can affect time. Light carries energy, and Einstein had shown that mass and energy are equivalent – so light should also be able to warp space and time.

Mallett's idea is to use an array of powerful lasers to create a circulating beam of intense light. If two pairs of parallel laser beams were set up, the space enclosed by the beams would be a square. And inside that square, Mallett says, empty space becomes "twisted" in much the same way that milk in a coffee cup begins to swirl when the coffee is stirred. The effect can be amplified, he says, by forcing the light to move in a spiral path, which he believes can be accomplished by means of something called "photonic crystal wave guides." (Photonic crystals are crystals whose optical properties allow light to propagate through them along specific paths.) If the beams of light are intense enough, the warping of space and time close to the beams could be severe enough to create a "loop" in time, Mallett says. In the jargon of physics, such a system would create a *closed timelike curve** – the key phenomenon at the heart of time travel. A particle that traveled along such a curve would, in theory, travel into its own past, just as walking around the block brings you back to the point at which you started. Mallett becomes more and more animated as he speaks; at one point he almost knocks a heavy textbook – appropriately titled *Gravitation* – onto the floor.

As crazy as Mallett's proposal may sound, he has described its key theoretical components in peer-reviewed physics journals. The reaction from the physics community, however, has been muted. The head of his department at the University of Connecticut, William Stwalley, has expressed an interest in the practical challenges of setting up Mallett's proposed experiment but cautions that actually making any sort of time machine "seems like a distant improbability." Two physicists at Tufts University recently wrote a critique of Mallett's theory, suggesting that any closed timelike curves that such an apparatus creates would occur at distances "greater than the radius of the visible universe by an immense factor."

Mallett remains hopeful, saying that it's not the physics or the math behind his proposal that experts have questioned, but rather the practicality of harnessing the equations to create a "time loop." Mallett has begun collaborating with an experimental physicist colleague, trying to see what it would take to get his laser-driven time loop up and running. (They're still in the process of trying to raise funds to test their ideas; Mallett believes his method could be tested conclusively with about $250,000 worth of equipment.) He expects to see the first results within a decade, and he believes that human time travel could be a reality before the end of the century.

---

* Mallett calls it a "closed timelike line," but the more common term is "curve."

Signaling in Seattle

Mallett is not alone in his quest. On the opposite coast, physicist John Cramer of the University of Washington is working on his own esoteric time-related contraption, a setup involving a modest array of lasers, mirrors, beam-splitters, and, most importantly, pairs of "entangled" photons. While Mallett has focused on relativity, Cramer is banking on that other pillar of modern physics, quantum theory – and, in particular, the strange idea of quantum entanglement (also known as "quantum non-locality"). First proposed in the early twentieth century, quantum entanglement suggests that observations of one member of an entangled pair of particles – making a note of its spin or its polarization, for example – would automatically yield information about the other member of the pair, even if they were separated by great distances. Even though it sounds quite strange – Einstein was very unhappy with the idea, dismissing it as "spooky action at a distance" – quantum entanglement has recently been confirmed by numerous experiments.

Scientists have weighed in with a variety of interpretations of quantum entanglement, trying to make sense of what it says about our universe. For Cramer, the simplest way to understand the phenomenon, with its apparently instantaneous communication between distant particles, is through an idea called *retrocausality*. As we sat in Cramer's Seattle office, the soft-spoken professor tried to explain retrocausality in layman's terms. It is, roughly, the peculiar state of affairs in which the future can affect the present or the present can affect the past – the subatomic equivalent of arriving at work before you've left the house. Though it sounds wildly counterintuitive, there's nothing explicit in the laws of physics that rules out such influence.

The tricky part will be demonstrating retrocausality in the laboratory. At the heart of Cramer's planned experiment is a beam of light that will pass through a crystal, splitting it into two beams of entangled photons. Each beam would then pass through a screen with a double slit – exactly the kind of setup that pioneers of quantum theory used nearly a century ago to show that light could behave as either a particle or a wave. After passing through the screens, both beams are focused onto detectors, though you would have no way of knowing in advance whether a given beam will register as waves or as a stream of particles. There's one more vital component: the detector on the second beam is movable; by adjusting its distance from the screen, you can control whether that beam is recorded as a wave or as particles. Because the beams are entangled, however, you know that if a particular photon at the second detector comes through as a wave, for example, then its entangled "partner" at the other detector will also appear as a wave. Cramer's bold idea is to force the second beam on a "detour" through some ten kilometers of coiled-up optical cable, slowing it down by a few microseconds. That means that your decision at the second detector – "I'd like to see a wave" or "I'd like to see some particles" – affects what's seen at the first detector, even though the beam has already hit the first detector a fraction of a second earlier.

At least, that's the theory. If the experiment works, future events could be thought of as influencing past events across that brief span of time. "In principle – if quantum nonlocality can be used to send signals – I can send a signal fifty microseconds back in time," he says. "I can receive it fifty microseconds before I send it." In principle, though, the delay could be much longer. "If you can have a fifty-microsecond difference between sending and receiving the signal, there's no reason, in principle, why you can't have a million of these, and have a fifty-second difference," he muses.

Like Mallett, Cramer has struggled to get funding for his planned experiment, though after his financial difficulties were mentioned in a local newspaper – the headline ran, "Physicist needs $20,000 for time-travel experiment" – generous donations began to pour in. In the article, Cramer lamented that not even the Defense Advanced Research Projects Agency (DARPA) would fund his experiment, even though, as the article pointed out, the agency was looking into "shape-shifting, liquid robots (think *Terminator 2*) as well as cyborg insects (half robot, half normal bug)." Cramer is glad to have the donations coming in, but, like many scientists, he's reluctant to spend too much time in the limelight. "I feel a little uncomfortable about having this publicity about an experiment that we haven't done yet," he says.

Once again, the magazine *New Scientist* was among the first to publicize Cramer's ideas. With the journal's usual flair, the writer declared that if retrocausality is confirmed – which he admitted was "a huge if" – it would "overturn our most cherished notions about the nature of cause and effect and how the universe works."

The response from the physics community has been more subdued. Even though quantum entanglement is by now well established, it is not at all clear whether one can use pairs of entangled particles for any kind of signaling, which is pretty much a prerequisite for Cramer's proposed experiment. In other words, entanglement only gets you partway to retrocausality.

Perhaps the biggest difference between Cramer's outlook and Mallett's is that Cramer expects his experiment to fail. He strongly suspects that this sort of backward-signaling can't occur. But exactly *why* it can't happen, he says, is a matter for investigation. "Very likely there's some relationship that will prevent you from doing this kind of signaling," Cramer says. "What we'd like to do is to use this experiment to understand what that relationship is . . . If you have the wherewithal to do the experiment, you ought to try it – try to push the envelope and see what happens."

Who hasn't dreamed of escaping time's prison? We seem to be caught in time's flow, moving inexorably forward one day after another. Imagine what it would be like if we had the same kind of freedom to move through time that we have to move through space: we could see a loved one who's no longer with us. We could peer into the future and see our great-great-grandchildren – along with stock market quotes from next month or next year. Or perhaps we would attempt more dramatic voyages: we could bear witness to the Crucifixion, or the Battle of Hastings (as countless writers have imagined). Or we might try to change history – to visit Berlin circa 1933, for example, and do away with Hitler before he could unleash his armies on the world.

It is little wonder that we find the subject of time travel so enticing. It has, of course, been a favorite topic for science-fiction writers for more than a century, from Wells's pioneering novel to the campy *Back to the Future* movie trilogy, the *Dr. Who* series, *The Terminator* and its sequels, and innumerable episodes of *Star Trek* (in all its incarnations).

Fiction Leads the Way

H.G. Wells was not the first writer to speculate on time travel – or at least, not the first to "play with time" by taking the reader across the years by clever narrative techniques. In Charles Dickens's *A Christmas Carol* (1843), Ebenezer Scrooge is whisked off to Christmases past, present, and future – but these could be seen more as visions than as physical journeys. In Mark Twain's *A Connecticut Yankee in King Arthur's Court* (1889), the main character receives a blow to the head and awakens to find himself in medieval England – though how he got there is never explained. Two lesser-known stories from about the same time deal with time travel more directly. Edward Page Mitchell's "The Clock That Went Backwards" (1881) and Lewis Carroll's "Sylvie and Bruno" (1889) both involve timepieces which, rather than merely *showing* the time, allow the user to *control* time, transporting him to whatever time is displayed.

But *The Time Machine* was different. For the first time, we were asked to imagine a purpose-built machine that gave the user complete control over time. The originality of this idea can hardly be overstated: the *Encyclopedia of Science Fiction* credits it with "a crucial breakthrough in narrative technology," an idea "so striking as to constitute a historical break" with what had come before.

For Wells, time was merely a dimension – one that could be traversed, just as we travel through space. *The Time Machine* was published in 1895, ten years before Einstein's special theory of relativity. Wells is best known for his writing, but he also took several science classes at London's Imperial College, and followed scientific developments with great interest. His vision clearly anticipated some of the key concepts that would underlie Einstein's theory.

In the wake of *The Time Machine* came a flood of time-travel stories of varying degrees of complexity and plausibility. Perhaps the most convoluted time-travel tale – and certainly one of the most disturbing – is Robert Heinlein's short story "All You Zombies –" (1959). In this twisted tale, a time traveler has a sex-change operation and ends up becoming *both* of his own parents. In fact, as the story progresses, we discover that all the major characters in the story are the same person, at different stages of his (or her) life.

In a more humorous vein, Douglas Adams tells us in the second instalment of his *Hitchhiker's Guide to the Galaxy* series that the main obstacle to time travel "is quite simply one of grammar," and that

> the main work to consult in this matter is Dr Dan Streetmentioner's *Time Traveller's Handbook of 1001 Tense Formations*. It will tell you for instance how to describe something that was about to happen to you in the past before you avoided it by time-jumping forward two days in order to avoid it. The event will be described differently according to whether you are talking about it from the standpoint of your own natural time, from a time in the further future, or a time in the further past and is further complicated by the possibility of conducting conversations whilst you are actually travelling from one time to another with the intention of becoming your own mother or father.
>
> Most readers get as far as the Future Semi-Conditionally Modified Subinverted Plagal Past Subjunctive Intentional before giving up: and in fact in later editions of the book all the pages beyond this point have been left blank to save on printing costs.

Must time travel of the sort imagined by these writers remain a fantasy, or is there any way it could be achieved in the real world? And, if so, how can we resolve the many paradoxes that such travels would seemingly entail?

Leaping Forward

Before we go any further, we must make a distinction between travel into the future and travel into the past. Thanks to the time dilation effects that arise with special relativity, traveling into the future is as simple as moving quickly for a period of time and then returning home. In fact, it's been done: the Apollo astronauts, as well as those who have spent a long time in Earth orbit, have returned to Earth having aged a tiny amount less than their stay-at-home colleagues (typically just a few milliseconds, mind you, because their speed was still a snail's pace compared with the speed of light). The current record holder for this sort of time travel is the Russian astronaut Sergei Krikalev, who spent more than eight hundred days whipping around the earth on board both the Mir space station and the International Space Station. So far he has aged about one-fiftieth of a second less than his colleagues on the ground.

The second part of Einstein's theory, general relativity, can also aid in the forward-time-travel effort. Spending time in the vicinity of strong gravitational fields – near a black hole, for example – also allows the intrepid traveler to age less than a stay-at-home sibling, again allowing for travel into the future (from the traveler's perspective).

I suspect, though, that many people may feel as if the sort of time travel experienced by Krikalev "doesn't count" – perhaps because the intervals are so small that the age difference isn't noticeable. But the size of that time difference is limited only by technology. In principle, an astronaut could embark on a long voyage and return to Earth to discover that many centuries had passed. Let's say you want to circumnavigate the Milky Way galaxy, a trip of roughly 150,000 light-years. Suppose you accelerate at a nice, low rate, increasing your speed by just 10 meters per second each second – this is "1g," which merely simulates the same force that gravity exerts on you every day here on Earth. Keep up that rate of acceleration for long enough and eventually, you will achieve enormous speeds. After a bit more than 11.5 years, you'll find that you've completed the first half of the journey. You'll have traversed 75,000 light-years, and you'll be whizzing along at 99.99867 per cent of the speed of light. Now begin decelerating at the same rate: after covering another 75,000 light-years, you're back home on Earth; you've completed one "lap" around the galaxy. But your clock is now wildly "off" from Earth's clocks: to you it will seem as though 23.16 years have passed. But on Earth, 150,002 years will have gone by.

This is not controversial: the time dilation effect of special relativity is established science. As physicist Brian Greene puts it, "There is every reason to believe, and no reason not to believe, that special relativity is correct, and its strategy for reaching the future would work as predicted. Technology, not physics, keeps it tethered to this epoch."

Yet this kind of time travel is disappointing in one significant way: it doesn't allow for a return journey. The astronaut can't see her future and then go back to the time that she departed from, bringing the news of, say, 2109 to the world of 2009 (along with a hundred years worth of stock prices, Super Bowl scores, and lottery numbers). That would involve not only forward time travel but also the more problematic backward time travel, which we'll get to in a moment. (For what it's worth, though, a forward-going time traveler, upon reaching her destination, would be perceived as a time traveler *from the past* – assuming people believe her story. With her intimate knowledge of twenty-first-century life, she would be of great interest to historians.)

Onward to Yesterday

Traveling into the past is more complex because of the apparent paradoxes it entails. The most famous problem is the so-called grandfather paradox, in which a time traveler kills her grandfather, thus preventing her own birth – and her journey back in time. There are many variations on this paradox, and we'll look at them more closely in a moment. But first it is worth noting that, oddly enough, there is nothing in the known laws of physics that specifically prohibits backward time travel. In fact, general relativity, with its remarkable warping of space and time, seems almost tailor-made to allow such peculiar journeys to occur. As physicist Lawrence Krauss has put it, "Einstein's equations of general relativity not only do not directly forbid such possibilities, they encourage them."

As we've seen, one of the crucial aspects of general relativity is the link it creates between matter and the geometry of spacetime. Matter literally warps the space and the time that surrounds it. So how do you achieve such a high degree of warping – intense enough to create a closed timelike curve? There is nothing in our solar system that even comes close to doing the trick. Instead, we need more unusual astrophysical objects – which brings us once again to the black hole, and its even more exotic cousin, the *wormhole*.

At first it was presumed that a body falling into a black hole was on a one-way journey: falling past the event horizon, such a body literally disappears from our universe and is soon crushed into oblivion at the center of the black hole. But then physicists began to consider another possibility: What if the black hole had an "exit" as well as an "entrance"? Such a double-ended black hole is known as a wormhole. In theory, a wormhole could act as a "bridge" across distant points in space and time, depositing a traveler into another universe – or perhaps a distant part of our own universe, depending on where (and when) the two ends of the wormhole are located.

The Worm Turns

Wormholes, however, come with a lot of baggage. For one thing, it has never been clear if such structures could be formed from normal matter – the stuff that stars and galaxies are made of. For starters, the amount of matter required to build a wormhole is enormous; would such an agglomeration not simply collapse under its own weight? In the 1980s, Kip Thorne, a physicist at the California Institute of Technology, suggested a solution: if the wormhole were made of an unusual sort of matter, capable of generating "negative energy," it could exert enough pressure to remain open. Such exotic forms of matter are suggested by quantum theory, but we have no way of knowing if wormholes made from negative energy are likely to exist in our universe. (It has been suggested that such structures may have been formed at the time of the big bang, and are perhaps now scattered throughout the cosmos. This is, however, quite speculative.) Even so, just the fact that such structures *could* exist – that they're not obviously prohibited by the laws of physics – was a stunning discovery. If a stable, navigable wormhole were ever found, an astronaut could enter in one location and emerge not only in another place but *at another time*, either in the past or the future. A wormhole would be a time machine. (There is one restriction on the sorts of journeys that could be undertaken, however: the traveler could not go back in time beyond the moment when the wormhole formed.)

It didn't take long for wormholes to find their way into science fiction. In the spring of 1985, Carl Sagan was revising his manuscript for his novel *Contact* and asked Thorne for his advice. Sagan needed a way for his heroine, Ellie Arroway, to traverse vast stretches of space and time. He had been thinking that a black hole might do the trick, but Thorne realized that a wormhole would be more appropriate. *Contact* was published later that year, and a popular movie adaptation, starring Jodie Foster, was released in 1997. Interestingly, as author David Toomey has pointed out, the inspiration flowed both ways: after advising Sagan, Thorne went on to consider the implications of wormholes in more detail. A paper published by Thorne and two colleagues, titled "Wormholes, Time Machines and the Weak Energy Condition," appeared in the journal *Physical Review Letters* in September 1988. Although just three pages long, it was one of the first papers by an established scientist to address time travel seriously. It was at this moment, Toomey writes, that "the idea of time machines crossed from the realm of science fiction into the realm of science."

Wormholes are probably the most widely accepted possible means of achieving time travel, but they're not the only method. Physicists have imagined other equally exotic structures that might also warp space and time to the required degree. In the 1970s, physicist Frank Tipler found that a long, rotating cylinder – if its surface was moving at close to the speed of light – would drag spacetime around with it, possibly creating closed timelike curves. (One hitch: the cylinder might have to be infinitely long.)

In the early 1990s, physicist J. Richard Gott envisioned a scenario employing "cosmic strings," long, dense formations of pure energy that may have been created in the big bang. Should two such strings happen to pass each other at high speed, an astronaut traveling around the pair at that moment could return to his starting position before he had left. A variety of other equally esoteric scenarios – all based on solutions to the equations of general relativity – have been put forward.

If backward time travel is somehow found to be possible, perhaps by means of a wormhole of the sort described by Thorne and his colleagues, we would be immediately confronted with the paradoxes associated with such travel. Those paradoxes raise many troubling philosophical issues, and it's worth a brief detour to examine a few of them before we proceed.

Philosophical Interlude, Part Two

Perhaps a good starting point is the question of the "reality" of the past and the future. The very idea of time travel presumes that there is a destination: that there is a past and a future, which are as real as the present. Not everyone agrees. One could object on the grounds that the past is "gone" – one cannot visit ancient Rome for the simple reason that ancient Rome does not exist. Perhaps one cannot witness future events for the same reason. (That line of reasoning obviously puts a damper on just about every time-travel scenario from science fiction: How can the Terminator robot, and an agent determined to foil its evil mission, arrive from a future that hasn't happened yet?) Both past and future time travel would be ruled out because there is nowhere – or "nowhen" – to go. Such a philosophical position – denying the reality of past and future – is called *presentism* (we looked at it briefly in Chapter 6), and it is occasionally put forward as a basis for the "no destination" objection to time travel. (Interestingly, it has occasionally been argued that presentism used to be the dominant world view – and that this explains the absence of time-travel stories prior to the middle of the nineteenth century.)

Another objection is called the "double occupancy" problem: How can I travel one second into either the past or future without colliding with the "version" of myself who is already there? Philosophers manage to write lengthy treatises on these sorts of issues, and some of it makes for fascinating reading. My impression, though, is that much of what is written stems (perhaps subconsciously) from a Newtonian perspective – as though time and space were quite distinct; as if there is one "master clock" for the universe. As we saw in Chapter 7, however, this is not the case: special relativity forces us to think of time and space as being intertwined. "Past" and "future" are on just as strong a footing as "the present," even though "now" is reduced to a subjective label. Indeed, Mr. Krikalev, the Russian astronaut, seems to have already overcome these obstacles. The alleged nonreality of the future apparently didn't prevent him from travelling one-fiftieth of a second into it; nor did he collide with the version of himself that was already there.

What, exactly, do we mean when we say "traveling back in time"? In general, we do *not* mean "living" or "experiencing" backward – whatever that might entail. (Although some authors have indeed examined that notion, often in excruciating detail – as in Martin Amis's novel *Time's Arrow*.) Most of us, even if we may wish to appear younger, would not want our memories, knowledge, and experience stripped away in the process. What we usually imagine when we speak of a time machine is something more complex: we imagine a machine in which we perceive the time "out

Calvin and Hobbes © 1988 Watterson. Reprinted by permission of Universal Press Syndicate. All rights reserved.

there" moving forward or backward at some rate that we can control, while our "local" time – inside the machine, presumably – runs along at the usual rate. (The philosopher David Lewis, who sparked a resurgence of interest in such matters with a 1976 paper titled "The Paradoxes of Time Travel," uses the terms "external time" and "personal time" to distinguish these two timelines.) This was the sort of time travel envisioned by Wells: when his time traveler pressed a lever to begin his journey forward in time, the maid "seemed to shoot across the room like a rocket," but inside the machine, time was clearly flowing normally. The time traveler does not experience accelerated aging, and if he had been wearing a watch, the implication is that it would continue to tick away as usual. Crucially, however, the dials on the machine's "dashboard" are somehow able to indicate the time "out there," successfully informing him when he has reached the year A.D. 802,701. (Fans of the original *Star Trek* series will recall an episode that reveals a similar dichotomy: in "The Naked Time," the *Enterprise* is sent hurtling back in time; while the crew's conscious experience appears to go forward, the ship's chronometer somehow records the "real time," and its wheels indicating the date and year can be seen to spin backward.)

There is also the thorny issue of whether one moves in space while one travels in time. It's not so much of an issue for the *Enterprise*, which can come and go as it pleases; but we can be sure that Wells intended his Earth-bound time machine to "ride along with the planet," so to speak. After all, the Earth is traveling in its orbit around the sun at some thirty kilometers per second, yet Wells's time traveler never gets "left behind" to materialize in outer space. At the very least, such issues force us to think very carefully about what we mean by "time travel" and what we would like our time machine to do.

Duck, Grandpa!

Let us turn now to those famous paradoxes. In the most-discussed scenario, a time traveler returns to, say, 1930, determined to kill his grandfather before he's had a chance to meet his future wife, the traveler's grandmother. (Or if we prefer, we can imagine that the traveler is simply a bad driver and runs over his grandfather by accident.) What would prevent such a seemingly impossible event from occurring?

There are four main kinds of solutions that have been put forward. One is that the laws of nature somehow conspire to prevent you from carrying out your nefarious mission. A second is that you can do as you please, but that since you've in a sense "already done it," the outcome will simply be what history has already recorded; if your grandfather lived to a ripe old age, you won't be able to change that fact. (As we will see, these first two solutions are essentially the same thing, viewed from a different perspective.) A third possibility is what we might call the "parallel universes" solution: in this scenario, you can travel into the past and kill your grandfather, but your actions do not change "history" – the history of the world as you knew it – but merely the history of another, parallel world. And finally there is the possibility that these various paradoxes serve as proof that time travel into the past is impossible.

Let's begin with the natural-conspiracy idea. You've gone back in time, emerging from your trusty time machine with only a slight headache. A newspaper confirms that it is 1930, and you track down Grandpa at the house where you know he lived at that time. Raising your pistol, you have him clearly within your sights. You can't miss. You squeeze the trigger, and ... what? This is where it gets interesting. Those who say that the laws of physics will prevent you from killing your grandfather are not suggesting that any sort of supernatural intervention takes place. Rather, they suggest that some mundane event – or series of mundane events – prevents you from carrying out your task. Perhaps in spite of those hours of practice at the shooting range, you miss. Maybe you have a change of heart, and decide at the last minute not to proceed. Perhaps you slip on a banana peel on the way to your grandfather's house. One way or another, a series of such events must keep you from carrying out your goal.

At first glance, such reasoning seems to be at odds with the notion of free will. (Actually, many philosophers – and quite a few physicists – are happy to declare free will to be an illusion, but let's assume, for now, that it's real.) If you're hell-bent on killing your grandfather, why would you "accidentally" keep slipping on banana peels or encountering other similar obstacles? Is some kind of "global" law interfering with what we do "locally"? If so, it would be a violation of what Oxford philosopher Michael Lockwood calls the *autonomy principle*. This principle states that we can do whatever we want locally "without reference to what the rest of the universe may be doing." In other words, the universe at large has no bearing (or shouldn't have any bearing) on what you do here and now, so long as you don't violate any physical laws locally. The universe may "require" that you fail to kill your grandfather – but why should the universe at large have any bearing on your ability to aim a gun at the poor fellow and pull the trigger, *here and now*? In a universe without any closed timelike curves, universal laws would never interfere with your local autonomy in the first place. But bring such time loops into the picture, Lockwood argues, and suddenly they place a sharp constraint on what you can do. When any one *particular* attempt to kill your grandfather fails, the explanation can be found locally – a banana peel or some such cause – but the reason for a *string* of failures "will lie in the global structure of the space-time manifold" in which we live. (One could argue, perhaps, that an endless series of "accidents" would seem just as bizarre – and in need of explanation – as the paradoxes they're supposed to circumvent; and, naturally, philosophers have attempted to deal with this objection at length.)

What's Done Is Done

The assumption here, which I think most physicists and philosophers would accept, is that you can't change history. As Brian Greene puts it, "If you time-travel to the past, you can't change it any more than you can change the value of pi." Certainly in special relativity, space and time simply *are*; what's done is done. Naturally, this acutely constrains the variety of actions you can carry out once you've traveled into the past. This brings us to the "you've already done it" argument, more formally called the *consistency principle*. This principle, expressed somewhat technically by the philosopher John Friedman, states that "the only solutions to the laws of physics that can occur locally in the real Universe are those which are globally self-consistent." Or, in layman's terms, if you travel back in time, you can only do things that are consistent with the overall history of the universe. Or, more simply yet: *If you travel into the past, you will only be able to do what you actually did.*

We see now how the first and second arguments amount to the same thing: Must the time traveler blame his failure to kill his grandfather on an endless series of puzzling coincidences, peculiar constraints, or a bizarre absence of free will? Such reasoning puts the cart before the horse. As Nicholas Smith, in his wryly titled paper "Bananas Enough for Time Travel?", says, "If a time traveler is going to travel to some past time, then she has already been there. If she is going to save a life or prevent a birth when she gets there, then she has already done so." A string of events can be labeled "improbable" only if we make some statement, in advance, about their likelihood; calling a series of events that has already happened "improbable" seems to be a misuse of the word. If Grandpa remembers a crazed stranger trying to kill him in 1920, then the time traveler simply is – was! – that stranger. As Smith argues, it is a mistake to think that some particular event happens, and then "happens again" when experienced by a time traveler. It simply happens, period. Smith refers to this mix-up as the "second-time-around" fallacy. If a time traveler witnesses some historical event, so be it; but that means she was there all along.

That idea sounds simple enough, but it still threatens to wreak havoc with the idea of free will. Consider, for example, the actions of Henry, the time-traveling hero in Audrey Niffenegger's *The Time Traveler's Wife* (2003). At one point, Henry, aged twenty-four, visits himself at age five. The older Henry struggles to make sure he does all the things he remembers seeing "himself" do, nineteen years earlier. Yet he seems to choose his actions freely. Given that his journey has "already happened," should he not simply be *compelled* to act precisely as he remembers seeing himself act? (Or perhaps he *is* compelled, and merely *feels* he has a choice . . . ?)

Billiard Balls and Bootstraps

Even simpler paradoxes can be imagined. Imagine a billiards table that just happens to have a carefully positioned wormhole on it. Now suppose one of the balls is heading toward the mouth of the wormhole. It enters the hole, then emerges from the other end just a moment before it had entered. We can imagine a scenario in which it strikes itself – perhaps altering its trajectory just enough so that it doesn't enter the wormhole after all. But if it doesn't enter the wormhole, then it can't have struck itself, therefore its path is not deflected, and so it enters the wormhole . . . and so on. (One rebuttal is that so long as the ball only strikes itself with a glancing blow – a blow not severe enough to prevent its entry into the wormhole – then everything is just fine. It's still a very peculiar case of an event causing itself – sometimes called the "bootstrap paradox" – but not a logical contradiction.)

Time-travel paradoxes need not involve people or even solid matter – they can involve information. There are many versions of this variation, often called the "knowledge paradox." They all go something like this: Suppose you've been struggling to solve a problem in physics or mathematics (it could be any field). For the sake of argument, let's say you're a math enthusiast, and you're curious about Goldbach's conjecture, a problem that has stumped mathematicians for more than 250 years. (The conjecture states that every even number can be expressed as the sum of two prime numbers: $4 = 2 + 2$; $6 = 3 + 3$; $8 = 3 + 5$ . . . Mathematicians strongly suspect that the conjecture is true – that the statement holds for all even numbers – but thus far there is no proof.) Let's say it's 2008, and you guess that mankind will need several decades to prove Goldbach's conjecture, so you set the controls on your time machine for the year 2040. Upon arrival, you head for the library (or the digital equivalent) and find to your delight that Goldbach's conjecture has recently been solved; you study the published solution with great interest. The proof is long and quite technical, but eventually you memorize it. You also recognize the author's name: though he is now a mathematics professor at a nearby university, back in 2008 he was a young child who lived in your neighborhood. You also glance at the publication date: 2038. Good. That will give the boy a solid thirty years to become a professor, come up with the proof, and write the paper. Returning to 2008, you give the child all the encouragement possible; he harbors doubts about his mathematical skills, but at your insistence he majors in mathematics, and eventually becomes a professor. But the years go by, and he seems to be making no progress at all on Goldbach's conjecture. He has published only a few research papers, none of them particularly noteworthy. Finally, with only a few days left before the publication is due to arrive at the journal's editorial office, you realize he isn't going to make it. Having memorized the proof, you write it up on some loose-leaf paper, slip it into an envelope, and slide it under the professor's door. Naturally, he is delighted to find this anonymous treasure: it will save his career and establish his reputation for all time. He promptly types it up, and submits it – right on schedule – to the mathematics journal where you had seen it.

So where did the proof come from? Not from the professor; he merely copied it from the material you slipped under his door. But it didn't come from you, either – you merely gleaned it from a mathematics journal! The knowledge appears to have come, quite literally, from nowhere. As with the closely related bootstrap paradox, this sort of "free lunch" seems bizarre, though it is not strictly speaking a logical contradiction.

Changing Whose History?

We turn now to the third possibility, the "parallel universes" solution. It quite successfully gets around the various paradoxes: if you kill your grandfather, for example, it merely means that there is a separate universe in which that event takes place. It is certainly a "tidy" theory – but what it gains in paradox-elimination one might say it makes up for in metaphysical baggage. Is there any reason to suspect that multiple universes actually exist?

In the strange world of quantum mechanics, the idea of multiple universes is not new. It can be traced (in its modern incarnation) back to the "many worlds" interpretation of quantum mechanics, put forward by physicist Hugh Everett III in the 1950s. In the many-worlds framework, every time there is an event with more than one possible outcome, *all* of the outcomes are said to occur – each in a different universe. Some critics might say the very idea of multiple universes flies in the face of Occam's Razor – the idea that explanations should be as simple as possible. Why postulate an infinite number of unseen universes when one will do? The reply, at least from those who support the idea, is that one universe *won't* do.

The many-worlds interpretation of quantum theory remains controversial, but a number of prominent theorists have come to embrace it in recent years. The most outspoken proponent of the idea is Oxford-based physicist David Deutsch. In his 1997 book *The Fabric of Reality*, Deutsch says that quantum theory, and in particular the many-worlds view, resolves the paradoxes associated with time travel. "The quantum theory of parallel universes is not the problem, it is the solution," he writes. "It is the explanation – the only one that is tenable – of a remarkable and counter-intuitive reality." When I spoke with him in his Oxford home, he was even clearer on the matter: "When you take into account quantum theory, time travel is indeed perfectly consistent," he says. "It's just that when you *do* go back into the past, you go back into the past of a different universe."

If you travel back and encounter your grandfather, this is a different (one might say "parallel") universe – one in which Grandpa *did* meet a mysterious traveler from the future. If you happen to kill him, it means he won't have grandchildren *in that universe*. In the universe you came from, however, he lives to old age; history (*that* history) is safe. The knowledge paradox is resolved in a similar way: in one universe, the professor discovers the proof on his own; in the other universe, the proof is copied from a paper published in that first universe. In neither case did the proof come "from nowhere."

Ronald Mallett, the Connecticut researcher trying to demonstrate time travel with beams of light, shares Deutsch's conviction in the reality of multiple worlds. "Our universe will not be affected by what you do in the past," he says with confidence. (He understands, of course, that were he somehow to save his father from an early death, it wouldn't change *his* universe; he would be saving him only in a new, parallel universe.)

Forbidden by Physics?

Finally, we have the fourth possibility: that the laws of physics somehow prohibit backward time travel. The most famous endorsement of this position has come from Cambridge physicist Stephen Hawking. He has dubbed such a universal law the "chronology protection conjecture": *The laws of physics conspire to prevent time travel by macroscopic objects.* Hawking admits that he doesn't know what the mechanism actually is that prevents such things – but whatever it is, it "makes the world safe for historians" by preventing any paradoxes from popping up. He also said, famously, that "the best evidence we have that time travel is not possible, and never will be, is that we have not been invaded by hordes of tourists from the future." But as David Toomey and others have pointed out, Hawking was clearly joking. After all, none of the time machines under consideration allow travel to a date prior to the machine's construction; the absence of time-traveling tourists, therefore, only means that no time machine has been invented *yet.* Even if an advanced civilization (perhaps *us* in a few thousand years) stumbled on a preexisting wormhole and was able to use it to travel arbitrarily far into the past, there are still several plausible reasons why we wouldn't see a plethora of time travelers. For example, couldn't the time-traveling tourists simply have worn the necessary disguises so as not to be noticed? (I imagine that anyone clever enough to spy on the assassination of Julius Caesar would also be clever enough to wear appropriate clothing so as not to stand out.) Another possibility is that we're one of the less interesting civilizations in the galaxy; as Deutsch and Lockwood write, future civilizations will have their own priorities, "and there is no reason to assume that visiting the earth [in this century] would be high on their list."

Interestingly, although Hawking today seems to support the chronology protection conjecture, he does have a record of wavering on the plausibility of time travel. Having ridiculed the idea in the 1980s, he then wrote in the introduction to Lawrence Krauss's *The Physics of Star Trek* (1995) that "One of the consequences of rapid interstellar travel would be that one could also travel back in time." He then told London's *Sunday Times* that, if one combines general relativity with quantum theory, time travel "does begin to seem a possibility." However, by the time he wrote his second popular book, *The Universe in a Nutshell* (2001), he again seemed to view time travel as highly improbable.

If one seeks "hordes of tourists from the future," perhaps it is prudent to explicitly invite them. That was the rationale behind a time-travel conference held at the Massachusetts Institute of Technology on May 7, 2005. Of course, it was also a forum for scientists to mull over the latest ideas about the science of time travel – but time travelers from the future were encouraged to drop by. (The idea for the conference was dreamed up by a graduate student named Amal Dorai; he had read in a comic book that only one such meeting was needed, because any time travelers from the future could attend, regardless of when the event was held.) As far as anyone knows, no such visitors showed up at MIT.

Each new paper on time travel, meanwhile, is greeted with the usual skepticism. Mallett's suggestion that human time travel could be a reality within a century draws its share of dismissive shrugs. Many researchers continue to focus on wormholes as the most promising method of time travel – but even wormholes have their detractors. In a fairly technical 2005 paper, physicist Leonard Susskind of Stanford University argues that there is no way to achieve a navigable wormhole without violating two seemingly sacrosanct laws of nature – the conservation of energy and the uncertainty principle of quantum mechanics. Potential time travelers "will need to find another kind of time machine," he says. Some physicists have also suggested that the equations of string theory prevent closed timelike curves from forming in the first place – whether via wormhole or any other mechanism. (We will take a closer look at string theory in the next chapter.)

And if time travel is shown to be impossible, that, too, would be a tremendously important discovery. As Hawking has put it, "Even if it turns out that time travel is impossible, it is important that we understand why it is impossible."

The answer may come from esoteric theoretical considerations by one of the giants in the field – someone like Stephen Hawking or Kip Thorne – or it may come from someone much further from the limelight, like Ronald Mallett or John Cramer, tinkering away in a basement laboratory with relatively modest equipment. Either way, the quest to understand time travel is about more than an infatuation with tales from science fiction. Thinking about time travel, with all its theoretical challenges and apparent contradictions, is forcing scientists to confront some of the most profound issues in modern physics: the nature of the universe's ultimate laws; deep questions about cause and effect; even fundamental questions about the nature of space and time.

Could it be just an absurd dream? Of course. When Mallett first encountered *The Time Machine*, he was eleven years old. Now he's older and wiser – and infinitely more grounded in the real world of physics. "It's our childlike wonder that gets squelched as we get older," he says. As a child, he admits, he was too young to grasp the difference between fantasy and reality. "For the child, that fantasy is real," he says. Yet somewhere inside, a glimmer of that childlike wonder remains – the wonder of reading about a bold inventor and a wonderful machine, and asking, "What if . . ."

IN THE BEGINNING
The search for the dawn of time

We aspire in vain to assign limits to the works of creation in *space*. . . . We are prepared, therefore, to find that in *time* also the confines of the universe lie beyond the reach of mortal men.

– Charles Lyell

In northeastern Arizona, the tributaries of a meandering river known as the Chinle Wash have chiseled out a fan-shaped trio of converging canyons, now protected as part of Canyon de Chelly National Monument. The Chinle Wash flows northward into the San Juan River, which in turn flows westward and empties into the Colorado; some two hundred kilometers to the west, these same waters have carved the gaping chasm of the Grand Canyon. Forged by the same river system, it is not surprising that Canyon de Chelly shares much of the same landscape and geology as its more famous cousin to the west. The trails that zigzag along the southern rim of the Canyon de Chelly system look out over sandstone cliffs that plunge as much as three hundred meters down to the desert floor below; mighty stone pillars like the spectacular Spider Rock leap back up to nearly as great a height. And while the term "off season" barely holds any meaning at the perpetually crowded Grand Canyon, Canyon de Chelly is mercifully quiet. On a recent spring evening, only one other traveler was enjoying the sunset at Spider Rock Overlook – a white-haired gentleman with an enormous 8x10 view camera on a massive wooden tripod, and, for good measure, a Hasselblad on a second tripod. (The equivalent scene at the Grand Canyon, I would discover a few days later, involves hundreds of camera-toting travelers perched shoulder-to-shoulder at just about every accessible spot along the popular South Rim.)

As the sun sinks toward the western horizon, the layers of rock on the northern and eastern canyon walls turn from pinkish brown to luminous orange and red. It is a visual feast – and it also allows one to peer across vast stretches of time. Canyon de Chelly's oldest feature is a layer of sandstone known as the Supai Formation, which formed 280 million years ago – tens of millions of years before the first dinosaurs walked the earth. Above that are the layers that give the cliff walls their warm, rose-colored hue. This is the so-called de Chelly sandstone – former sand dunes, now compressed into solid rock, dating from 250 to 230 million years ago. Some 50 million years later, the uppermost layer, known as the Chinle Formation, was laid down. Of that, only the base layer, known as the Shinarump Conglomerate, remains; the rest has succumbed to millions of years of weathering and erosion.

The rocks at Canyon de Chelly tell a remarkable tale, and time is the central theme in their story. But it has not been easy to decipher the message. It is only in the last few hundred years that we have come to recognize the vast scope of Earth's history, and a parallel discovery – that the entire universe is much older even than these layers of sandstone – is newer still. In ancient times, the vast reach of cosmic history lay hidden.

As we saw in Chapter 4, many ancient cultures envisioned time as a series of endless cycles; in such societies, the question of how or when time began would have been meaningless. "Two thousand years ago," writer Martin Gorst says in his entertaining book *Measuring Eternity*, "the idea that the world might have a starting point was inconceivable." But the idea of a single creation event, first conceived by the Jews (likely borrowing from the Babylonians) and later adopted by the Christians, suddenly made the search for origins a logical quest. For centuries the answer was believed to lie in scripture, rather than in nature – but the search had begun. Great thinkers like Augustine pored over the Book of Genesis, adding up the ages of Adam's male descendants – counting all those "begats" – so as to arrive at the time of the Creation. Augustine ended up with a date of 5500 B.C.; the Venerable Bede, a medieval English monk, had placed it slightly later, at 5199 B.C. Work of this sort continued for a thousand years, into the beginning of the modern era. Kepler, Newton, and Martin Luther all came up with dates in the rather narrow range of 4000 B.C. to 3993 B.C. (The traditional Jewish date of creation is 3760 B.C., although the Hebrew calendar takes as its starting point the previous year, 3761 B.C.)

The Bishop and the Bible

But the most influential statement on the age of the world came from an Irish bishop named James Ussher (1580–1655). Born in Dublin, Ussher was raised as a Protestant in a largely Catholic nation. He had a voracious appetite for books, and read everything in sight; he toured all of the great libraries of Britain and Ireland, and eventually amassed some ten thousand volumes, one of the largest private collections in all of Europe. Ussher looked not only at the Bible but at hundreds of other ancient texts, trying to align the different (and often conflicting) histories that they presented. He eventually pinned down the death of King Nebuchadnezzar II of Babylon to 562 B.C. – a date still accepted by historians. He then went through the Hebrew Bible (the Old Testament), adding up the ages of the prophets and the reigns of the kings until he arrived at Nebuchadnezzar, an interval of 3,442 years. Simple arithmetic then puts the Creation at 4004 B.C. Ussher would go even further, and fix the exact time and date. The apples were ripe in the Garden of Eden, he reasoned, so it must have been fall – likely, he imagined, the autumnal equinox. And since the Genesis account suggests each day began in the evening – "And the evening and the morning were the first day" – he believed the world was created in the evening. He eventually concluded that the world began at 6:00 p.m. on Saturday, October 22, 4004 B.C. Later

Irish bishop James Ussher. His study
of biblical chronology led him to
believe the world was created on
Saturday, October 22, 4004 B.C.
(National Portrait Gallery, London)

commentators usually omitted the extra detail, and the first full day,
October 23, 4004 B.C., became known as "Ussher's date."

Today we wonder how Ussher could have been so confident in a
date that would turn out to be so wrong, and that in any case seems sus-
piciously precise. But in Ussher's time, many scholars were performing
such calculations and obtaining similar results. Most of these studies were
forgotten, but the work of this otherwise obscure Irish cleric would
become household knowledge. Soon after he wrote his analysis, a London
publisher started printing bibles with Ussher's chronology in the margin;
in 1701, the Church of England endorsed his timeline in a new translation
of the Bible. (As the centuries passed, many readers simply assumed the
chronology was *part* of the Bible.) As late as the first decades of the twentieth century, bibles with Ussher's chronology were still being printed.

In the eighteenth century, however, some scholars were beginning to doubt
that the earth could be as young as Ussher had proclaimed. A few bold
writers openly suggested that Genesis was not an accurate description of

how the world began – or, at the very least, that it should be read metaphorically rather than literally.

Among these more adventurous thinkers was a Frenchman named Georges-Louis Leclerc de Buffon (1707–88). In his ambitious *Histoire Naturelle*, a forty-four-volume work that attempted to cover all of the natural sciences, Buffon argued that the earth was created by a comet hitting the sun. The idea made religious thinkers furious. "In a stroke, he had reduced the creation of the world, the glorious masterpiece of the Supreme Architect, to nothing more than a catastrophic accident," writes Martin Gorst. Buffon also suggested that all species were linked by inter-mediate gradations, and may have developed from a common ancestor; he believed that God did not concern himself with the details of each plant and animal. As for the biblical account, Buffon said that the language of Genesis had to be carefully interpreted. Echoing the words of Galileo, he said that the Bible had been written not for scientists but for lay readers. There was no need to presume that each "day" mentioned in the opening chapters of Genesis was of the same duration as a modern day of twenty-four hours. Indeed, the very idea of a "day" as a succession of day and night was established only after the third so-called day, with the creation of the sun.

Buffon instead took a scientific approach to calculating the age of the earth. He believed that our planet began as a sphere of molten rock and then gradually cooled down to its present temperature. He began to experiment with balls of different materials, heating them up and then observing how long they took to cool down. After six years of experi-mentation, he concluded that the earth was 74,832 years old.* He would later work out revised estimates, pointing toward an even more ancient Earth – but these he kept to himself. Perhaps he was taken aback by the large figures he was coming up with and feared that the public would not welcome such information. "Why does the human mind seem to lose itself

---

* Like Ussher's date, Buffon's estimate seems overly precise. He does not seem to have worried about his "margin of error."

in the length of time," he wondered. "Is it not that being accustomed to our short existence we consider one hundred years a long time, have difficulties forming an idea of one thousand, cannot even imagine ten thousand years, or even conceive of one hundred thousand years?"

Buffon had based his method on Newton's work on the physics of cooling objects, and Newton had, in fact, used the same technique to estimate the age of the earth. He had come up with a figure comparable to Buffon's – about fifty thousand years. But unlike Buffon, Newton could not accept the figure; it conflicted too sharply with his religious convictions.

Secrets in Stone

Even in the ancient world, there were some who dared to imagine an evolving Earth, a world whose appearance changed over time. Herodotus, for example, imagined geological processes taking many thousands of years. Around the year A.D. 1000, the Persian philosopher and scientist Avicenna imagined an ancient Earth in terms that sound remarkably modern: mountains, he said, "are effects of upheavals of the crust of the earth, such as might occur during a violent earthquake, or they are the effect of water, which, cutting for itself a new route, has denuded the valleys, the strata being of different kinds, some soft, some hard. The winds and water disintegrate the one, but leave the other intact." He concluded that it "would require a long period of time for all such changes to be accomplished."

In the century before Buffon grappled with the age of the earth, the English naturalist John Ray (1627–1705) was examining fossils he had uncovered in the Midlands and northern Wales; some of them, he could see, were of plants and animals that no longer existed. Would it not have taken thousands of years, he wondered, for such species to flourish and then become extinct? If so, it seemed to raise troubling theological questions: If God created a perfect world, why would some creatures die off, to be replaced by others?

Building on Ray's work, the geologist James Hutton (1726–97) traveled extensively throughout Britain and reached similar conclusions about the age of the earth. He believed that heat from the earth's interior occasionally pushed molten rock into the planet's crust, and that rocks such as granite were once molten. Such processes, he reasoned, required enormous amounts of time. "The result, therefore, of our present enquiry," he concluded, "is that we find no vestige of a beginning, and no prospect of an end."

For some, this new time scale was impossible to grasp; at the very least, it was difficult to reconcile such vast stretches of time with those mentioned in scripture. But at least one prominent philosopher was ready to embrace this new, longer view. Immanuel Kant (1724–1804) imagined that creation was "not the work of a moment" but rather an ongoing process. In words that seem to presage those of modern cosmologists, he writes, "Millions and whole myriads of millions of centuries will flow on, during which always new Worlds and systems of Worlds will be formed, one after another ... It needs nothing less than an Eternity to animate the whole boundless range of the infinite extension of Space and Worlds, without number and without end." As writers Stephen Toulmin and June Goodfield put it in *The Discovery of Time*, "By 1750, men could contemplate a future lasting many thousands of years; but no one before Kant had talked so publicly and seriously of a *past* comprising 'millions of years and centuries.'"

A century later, the geologist Charles Lyell (1797–1875) would conclude that the processes that shaped the earth in the ancient past were the same forces that continue to operate today. Change could come gradually, over thousands of years; cataclysmic events, such as Noah's flood, were unnecessary. Based on the fossil evidence, he concluded that even climates could change over time: Europe, he believed, was once a mild, tropical land. In his *Principles of Geology*, published in three volumes between 1830 and 1833, Lyell reasoned that the earth might be not merely thousands of years old, but perhaps *millions* – a startling claim. It is little wonder that when a colleague gave a copy of the first volume to a young naturalist named Charles Darwin, he urged him to enjoy Lyell's lively writing but to take his conclusions with a grain of salt. Even so, it must have made for provocative reading as the *Beagle* set off on its lengthy voyage.

Charles Lyell (left) was one of the first to suggest that geological processes unfolded over millions of years. Charles Darwin (right) took the first volume of Lyell's *Principles of Geology* with him as he set sail on the *Beagle*. *(Left: Dr. Jeremy Burgess / SPL / Publiphoto; right: SPL / Publiphoto)*

At around the same time, a contemporary of Lyell named George Scrope (1797–1876) was touring France, examining extinct volcanoes. He concluded that they formed gradually, by ongoing geological processes. His conclusion, published in his *Memoir on the Geology of Central France* (1827), is perhaps the most famous utterance ever delivered by a geologist: "The sound which, to the student of Nature, seems continually echoed from every part of her works, is – Time! – Time! – Time!"

Darwin's Deep Time

Charles Darwin (1809–82) was fresh out of university and thinking about training for the priesthood when one of his teachers suggested he take passage on the HMS *Beagle* as the ship's naturalist. The aim of the voyage was to conduct hydrographic surveys of the coast of South America, as well as the islands of the South Atlantic and eastern Pacific. The *Beagle* set sail in 1831 on a journey that was supposed to last two years but that in the end lasted five. On the long voyage, Darwin saw ample evidence of geological forces at work; in just the first few months, he was converted to Lyell's view of an ancient Earth. Darwin observed coral atolls, which he deduced were formed by volcanoes now vanished. In the Cape Verde Islands, he saw cliffs apparently thrust upward by a succession of volcanic events. In Chile, he witnessed an earthquake. He reasoned that the Andes themselves were formed by gradual geological upheaval. It was as though Lyell, in spirit, was on the deck of the *Beagle* alongside Darwin. The geologist's book "altered the whole tone of one's mind," he would later write. "When seeing a thing never seen by Lyell, one yet saw it partially through his eyes."

And then there were the animals. Darwin did not immediately know what to make of the great variety of species that he saw. On the Galapagos Islands alone there were dozens of different species of finch, each differing from the others only by slight variations in the size and shape of their beaks. Fossils, meanwhile, showed that species that had once flourished were now extinct, while new species had come to take their place. Here he would come to disagree with Lyell, who held to the notion that each species, once created, never changed.

Darwin's ideas about evolution began to take shape only after his return to England in 1836. After living briefly in London, he moved to a large country home southeast of the city. He married his first cousin, Emma Wedgwood (of the porcelain dynasty), and fathered ten children, seven of whom survived to adulthood. For the next two decades, he pored over his notebooks, studied his specimens, and considered what he had seen on his voyage. (At the same time he was battling a crippling illness, suffering constantly from vomiting, shivering, palpitations, and headaches. The illness was never diagnosed. Whatever it was, it often left him too sick to work; on his good days he was lucky to work for even a few hours.)

Darwin was not the first to speculate about evolution; in fact, his grandfather, Erasmus Darwin, had suggested that all warm-blooded animals had arisen from a common ancestor. But the question of *how* species changed over time was still unresolved. Charles Darwin's break-through was not the idea of evolution itself but the mechanism behind it: *natural selection*. (Roughly put: organisms that are best adapted to their particular environments are more likely to survive and reproduce, passing on whatever traits gave them an "edge" to their offspring.) This mechanism of natural selection was the tool that nature used to create new species of plants and animals. Evolution demanded time – and Darwin believed that, thanks to Lyell's geology, there was more than enough to go around. Darwin did not discover "deep time," but, standing on the shoulders of the geologists, he wholeheartedly embraced it. "What an infinite number of generations which the mind cannot grasp," Darwin wrote, "must have succeeded each other in the long roll of years."

On the Origin of Species was published in November 1859. The first edition sold out immediately, and ran through six more editions in the next dozen years. The scientific community was swift to embrace Darwin's ideas. There were, of course, theological objections: Darwin's theory suggested that all living things were related – he would later write that men and apes are essentially cousins – and because all species evolve gradually, no special "act of creation" was required.* These were difficult ideas to get used to; some conservative religious thinkers were never able to accept them. And looming over the entire debate was the vast evolutionary timeline. Ussher's chronology, with its six-thousand-year age for the earth, began to seem quaint. Our planet's history, and life itself, now seemed irrevocably to stretch back millions of years. "Darwinism," as writer Timothy Ferris has put it, "was a *time* bomb."

The impact of this revolution in our conception of time, which had gained almost universal acceptance by the end of the nineteenth century, was eloquently expressed by the geologist Archibald Geike: "How vast a period must have been required for that marvelous scheme of organic development which is chronicled in the rocks!" he declared in 1892. "The law of evolution is written as legibly on the landscapes of the earth as on

---

* Darwin would address the issue of human origins directly in his later book, *The Descent of Man*, published in 1871.

any other page of the book of Nature." That evolution is revealed both in the living world and in the planet itself:

> The living plants and animals of to-day have been discovered to be eloquent of ancient geographical features that have long since vanished . . . They tell us that climates have changed; that islands have been disjoined from continents; that oceans once united have been divided from each other, or once separate have now been joined . . . The present and past are thus linked together, into one vast system of continuous progression.

Darwin himself was awed by the ageless drama of evolution – and the vast epochs of time it demanded. As he wrote in *Origin*, "He who . . . does not admit how vast have been the past periods of time may at once close this volume."

The conclusions reached by the geologists and naturalists would soon be shared by the physicists. The discovery of X-rays in 1895 and of radioactivity a year later would open up a new world within the atom – and yield new tools for probing vast stretches of time. The New Zealand–born chemist Ernest Rutherford (1871–1937), working in Montreal, discovered that certain elements were unstable and released energy at an ever-decreasing (but predictable) rate, for prolonged periods of time. Minerals that contained those elements could release energy for thousands, even millions, of years. Rutherford immediately recognized the importance of his discovery: the earth's interior isn't simply getting cooler over time, as Buffon had assumed two hundred years earlier; instead, it is being continuously heated from within, by the decay of radioactive elements in the planet's molten core. Radioactivity "thus increases the possible limit of the duration of life on the planet," Rutherford wrote, "and allows the time claimed by the geologist and biologist for the processes of evolution."

The concept of radioactive decay would lead to a powerful new technique for determining the age of minerals. The rates at which these unstable particles decay are independent of changes in temperature and pressure, depending only on the particular element involved. Before long, geologists were using the new method of "radiometric dating" to measure the ages of rocks now revealed to be hundreds of millions of years old. In his book *The Age of the Earth* (1927), physicist Arthur Holmes declared, "All the evidence is consistently in harmony with the conclusion . . . that the age of the earth is between 1,600 and 3,000 million years" – that is, 1.6 to 3 billion years.* The age suggested by Ussher, and by now usually associated directly with the Bible, suddenly seemed like a drop in the vast ocean of geological time. "For a public used to dealing in everyday numbers of tens and hundreds," writes Martin Gorst, "this huge jump, from millions to billions, was dazzling."

Dazzling, yes – but fortunately for scientists, such numbers are no harder to manipulate than those one can count with one's fingers. We may not be able to picture these large numbers – it is hard enough to "see" fifty thousand fans in a packed sports stadium – but that does not get in the way of using such figures. Arthur Eddington, the British astronomer – he led one of the eclipse expeditions that confirmed Einstein's general relativity in 1919 – understood this paradox of large numbers. In a popular essay, he tabulated, in kilometers, the distances to the sun, the nearest star, and various other astronomical objects, out to the furthest galaxy known at the time, given as 3,000,000,000,000,000,000,000,000. "Some people complain that they cannot realize these figures," he writes. "Of course they cannot. But that is the last thing one wants to do with big numbers – to 'realize' them. In a few weeks time our finance minister in England will be presenting his annual budget of about £900,000,000" – this was in the days when national budgets were *less* than a billion pounds or dollars. "Do you suppose that by way of preparation, he throws himself into a state of trance in which he can visualize the vast pile of coins or notes or commodities that it represents? I am quite sure that he cannot 'realize' £900,000,000.

---

* The modern value is about 4.5 billion years, based on the oldest known rocks, meteorites, and lunar samples.

But he can spend it." Such large numbers "are not meant to be gaped at," Eddington says, "but to be manipulated and used."

Time and the Cosmos

The stars, too, tell a story that revolves around time – and this story, like the story of Earth's ancient origins, took centuries to interpret. While every culture had myths about how the universe began, the scientific study of cosmology is actually one of the youngest of the sciences; in its modern form, it began only in the first decades of the twentieth century. Although Galileo had seen that the fuzzy band of light called the Milky Way is really a vast collection of stars – far too many for him to count – it took more powerful telescopes, and more sophisticated new techniques for determining the distances to the stars, to reveal our galaxy's true structure. We now know that it is shaped like a giant Frisbee, some 100,000 light-years across,* with a bulge in the middle and spiral arms sweeping out to the edges. (From our perspective, embedded inside one of the spiral arms, it appears as the milky band that we see at night.) With the advent of larger telescopes, astronomers began to catalog other fuzzy patches of light, known as nebulae, seen scattered across the heavens. Some of them were observed to have distinctive spiral shapes; these at first were known as "spiral nebulae," but by the 1920s it was clear that they were galaxies like our own. These "external" galaxies were similar to the Milky Way, but incredibly distant. (In fact, Kant had suggested as much in the 1750s, referring to such nebulae as "island universes.") Our picture of the cosmos was growing.

But that was just the beginning. Astronomers were learning how to measure the distances to those galaxies, and by looking at their spectra, they could calculate the speed at which they were moving through space. The spectral lines of many galaxies were found to display a shift toward the

---

* A light-year is the distance that light travels in one year – roughly 9.5 trillion kilometers.

red end of the spectrum – a *redshift* – suggesting they were moving away from our own galaxy. (The phenomenon is analogous to the "Doppler effect" that makes the siren of a receding ambulance sound lower in pitch than it does when stationary.) Then, in 1929, the American astronomer Edwin Hubble (1889–1953) made a remarkable discovery. Hubble had been using the hundred-inch telescope at Mount Wilson in California – at that time, the largest telescope in the world – to systematically study distant galaxies, which he found "scattered through space as far as telescopes can penetrate." To his great surprise, he found a correlation between a galaxy's distance from the Milky Way and its motion: the farther the galaxy, the faster it was speeding away. He had discovered the expansion of the universe.

Hubble's new picture of the cosmos was a radical one. Until then, there had been no reason to imagine that the universe was anything but *static* – that it had always been there, and that it had always looked pretty much the way it looks now. The new picture was far more dynamic. We live in an evolving universe.

U.S. astronomer Edwin Hubble.
His study of galaxies beyond
the Milky Way showed that
we live in an expanding universe.
(Hale Observatories / SPL / Publiphoto)

Interestingly, the expansion of the universe was almost anticipated by Einstein. Just a few years after he worked out his equations for general relativity, he began applying them to the universe at large. But he was surprised by what they implied: the equations did not seem to allow a static universe; instead they appeared to demand that the universe must be either expanding or contracting. A handful of scientists were willing to explore such scenarios; the Russian cosmologist Alexander Friedmann, for example, found solutions to Einstein's equations that suggested an expanding universe, which he considered a perfectly reasonable scenario. A Belgian physicist named Georges Lemaître, who was also a Catholic priest, went further; even before Hubble announced his finding, Lemaître suggested that the observed redshifts could be a sign of the universe's expansion, and put forward the idea that the world began as a "primeval atom."*

Einstein would have none of it. Like the vast majority of scientists at the time, he believed in an unchanging universe. He went so far as to introduce a fudge factor into his equations, known as the *cosmological constant*, to "balance" the universe. Just a few years later, however, came Hubble's monumental discovery – and Einstein immediately regretted his cosmic fudge. He called it the "greatest blunder" of his career.

The Expanding Universe

The implications of an expanding universe are staggering. Wind back the clock, and the universe gets smaller. If you go back far enough, it must have been incredibly tiny. Before long, the notion of a cosmic explosion had taken root: the universe must have started off as an incredibly dense, hot fireball. As it expanded, it cooled. (Today, we're lucky to be living

---

* Although Hubble is generally credited with discovering the expansion of the universe, several other scientists, including astronomers Vesto Slipher and Milton Humason, made important contributions. The work of Lemaître was largely ignored at the time, in part because he published in a relatively obscure Belgian journal. Many now consider Lemaître the "father of the big bang."

Belgian priest and physicist Georges LeMaître (left), one of the originators of the "big bang" model of cosmology. Albert Einstein at first could not accept the idea of an evolving universe. *(Caltech Archives)*

near a nice warm star, because the average temperature of the universe is just a few degrees above absolute zero – more than 270 degrees below zero on the Celsius scale.) As the universe cooled, the first structures formed, leading, after billions of years, to the universe we see today, peppered with galaxies and clusters of galaxies. Astronomers call this the *big bang* model of the universe.* (The name was coined by the astronomer

---

* Although I've used the word "explosion" in describing the big bang, the term can be misleading. Physicists think of the big bang as an expansion of space itself, rather than an expansion of matter *into* a preexisting space. For this reason, there is no one "place" where the big bang happened, nor must the universe have an edge. Nonetheless, the universe may be finite, just as the surface of an expanding balloon is finite.

Fred Hoyle during a 1950 BBC radio broadcast. Ironically, Hoyle rejected the theory in favor of his own ill-fated "steady state" model.)

The work of Hubble and Einstein led to a new world view. The universe, it seemed, now had a beginning. (To some scientists, the big bang model smacked of Christian theology – and it didn't help that one of its originators was an ordained priest.) It now appeared as though time itself began at a precise moment, and, suddenly, it seemed within the bounds of science to determine when that momentous event happened. The quest for the beginning of time, which had started with the rocks underfoot, had moved into the heavens.

By the middle of the twentieth century, the evidence for the big bang was beginning to mount. Physicists learned how to measure the abundance of the chemical elements that make up the stars and galaxies, and it turned out that the amount of hydrogen, helium, and heavier elements was just what the big bang model predicted; the numbers were consistent with the notion of an expanding and cooling cosmic fireball. The icing on the cake, however, came in the mid-1960s. The clearest "fingerprint" of the big bang turned out to be a kind of echo of that initial explosion – not an echo of sound, but an echo of microwave radiation, discovered by accident in 1965 by two scientists using a giant radio antenna in New Jersey. Arno Penzias and Robert Wilson detected a faint microwave signal coming from every direction in the sky. They had discovered the signature of the big bang, now known as the "cosmic microwave background," or CMB. (By coincidence, at around the same time, physicist Robert Dicke and his colleagues at Princeton University – just an hour away – predicted the existence of the CMB based on theoretical models of the big bang. They were preparing to search for such radiation when they heard of the discovery by Penzias and Wilson.) One can think of the CMB as a splash of radiation that was released when the universe was less than half a million years old; today this radiation fills all of space. Penzias and Wilson received the Nobel Prize in 1978 for their discovery.

More recently, scientists have used a satellite known as WMAP – the Wilkinson Microwave Anisotropy Probe, launched in 2001 – to examine the background radiation in detail. With the data from WMAP, astronomers have been able to describe the basic features of the universe with remarkable precision. We now know, for example, that the universe is "flat," meaning that it can be described by the simple geometry of Euclid: parallel lines remain parallel, and the three angles of a triangle add up to 180 degrees. We also know that "ordinary matter" – stars and planets, for example – accounts for just 4 per cent of the contents of the universe. (Of the remainder, mysterious "dark matter" makes up about 23 per cent, and the even more mysterious "dark energy" accounts for 73 per cent. We will hear more about dark energy in the next chapter.) Finally, WMAP has let scientists pin down the age of the universe: astronomers now believe the big bang happened 13.7 billion years ago – give or take a couple of hundred million years.

The Ultimate Free Lunch

The discovery of the cosmic microwave background was an enormous breakthrough, but it also raised new questions. For one thing, the universe is awfully big – much bigger than physicists would have expected based solely on the big bang model. Also, the radiation was very "smooth" – that is, astronomers measured exactly the same intensity of microwave radiation no matter which direction they aimed their telescopes. (No part of the cosmic background radiation deviates from the average by more than a few parts per million.) If the universe were small, that wouldn't be so surprising: information could have passed back and forth from one part to another, smoothing everything out. But the universe is big – and it's only been around for 14 billion years. There just hasn't been enough *time* for this smoothing out to happen, across such large distances.

A tentative solution came in the early 1980s, when cosmologists came up with a modified version of the big bang picture, known as "inflation." According to inflation, the universe went through a period of incredibly rapid expansion during its first split-second of existence. This period of rapid (or "exponential") growth helped to explain why the universe is so large and so smooth. Several scientists worked on inflation, but the first crucial paper was written by physicist Alan Guth (now at MIT) in 1981. Guth's model doesn't specify exactly what was driving the inflation process. Physicists call the responsible entity a "scalar field" because it can be described by a single number attached to every location in space, rather like temperature – though this hardly helps us picture what is doing all of that pushing. But the inflation model appears to work: it successfully describes the subsequent evolution of the universe.

Inflation, incidentally, also gives us another dose of Copernican-style humbling: it suggests that the portion of the universe we can see with our telescopes – vast as it is – is just a small fraction of the larger cosmos. The visible universe (sometimes referred to as our "bubble" or "horizon") is surrounded by more distant regions – but these have been pushed away so rapidly by inflation in the universe's earliest moments that light from such regions has not had time to reach us. These far-off realms may remain forever unknown to us, as there is no way to receive any kind of signal from them.

Scientists are not used to the idea of something being created from nothing; even Aristotle shuddered at the idea, believing it an affront to reason. And yet the inflationary universe seems to entail just that. If the theory is correct, writes Alan Guth, "then the inflationary mechanism is responsible for the creation of essentially all the matter and energy in the universe." The origin of matter, he says, is no longer beyond the reach of science. "Conceivably, everything can be created from nothing . . . In the context of inflationary cosmology, it is fair to say that the universe is the ultimate free lunch." Inflation quickly became the "new and improved" big bang; it is now considered a standard part of the description of cosmic evolution. And, remarkably, all of the observations that astronomers have carried out so far appear to support the theory. (For example, inflation predicts the universe to be geometrically "flat," as WMAP has confirmed.)

At a recent conference on inflationary cosmology held in Davis, California, I asked Guth – mild-mannered, friendly, and eager to discuss his work – to give me the beginner's guide to inflationary cosmology. "Inflation is a twist on the big bang theory," he explained. "Inflation is an answer to the question, 'What caused the universe to propel? What triggered this gigantic expansion, that we're still seeing the universe undergoing?'" According to inflation, Guth explains, the scalar field exerted an enormous pressure on all of the matter in the universe. The universe ballooned in size, doubling in diameter again and again, a hundred times over. Before inflation, the universe was more than a billion billion times smaller than a proton; after inflation, it was about the size of a marble (or perhaps a grapefruit) – and it all happened in a tiny fraction of a second. With a steady stream of observational data from WMAP and other studies of the CMB appearing to support inflation, Guth says he has just as much confidence in the theory today as he did in 1981. "I think the observations certainly show that either inflation is right, or something that's very close to inflation is right, and explains how the universe got to be the way it is."

That is not quite the end of the story: cosmologists have come up with other scenarios that challenge the inflation model, and we will look at those shortly. Our picture of the origin of the universe will surely continue to evolve as bigger telescopes and more sophisticated theories reach deeper into the cosmos than we can peer at present. Our generation is not the first to believe it has nailed down the story of how the universe began, or even to pinpoint its age. As Martin Gorst reminds us, Ussher was not the only one who came up with the wrong answer. "Many of the greatest minds in science were equally blinkered, trapped by their own beliefs, or the prevailing assumptions of their day." Newton failed to take the idea of an ancient Earth seriously, and for many years Einstein balked at the idea of an expanding universe. Even geniuses can be led astray.

In the medieval world, people imagined a relatively short cosmic time scale, compatible with the view that the universe was made *for us* – a view that persisted well into the beginning of the modern era. The much longer timeline revealed by geology and later by cosmology was humbling; human beings appeared to play a much less central role in cosmic affairs. As Timothy Ferris has put it, "The larger the universe looms, the sillier it becomes to maintain that it was all put together for us." Nor is our location particularly special: we live on a typical rocky planet orbiting an average yellow star that lies partway out along a spiral arm in a very ordinary galaxy that we call the Milky Way. But from this outpost we have learned how to observe the greater cosmos, and for the first time we have data – real data from telescopes on the ground and in space – that reveal the structure and history of our universe with remarkable clarity. In that respect, modern cosmology is not "just another creation story," as some postmodernist scholars suggest. There are, of course, many more details to be worked out, but in its broad outline, we can be confident in the big bang model of the evolving universe. The universe as we know it – and time as we know it – began 13.7 billion years ago.

BEYOND THE BIG BANG

The frontiers of physics and the origin of time's arrow

Our picture of physical reality, particularly in relation
to the nature of *time*, is due for a grand shake-up –
even greater, perhaps, than that which has already
been provided by present-day relativity and quantum
mechanics.

– Roger Penrose, *The Emperor's New Mind*

The big bang model of our cosmic origins must be counted as one of the
great triumphs of twentieth-century science. In its latest incarnation –
the theory of cosmic inflation – it allows us to grasp a period beginning
a billionth of a trillionth of a trillionth of a second after the big bang (or,
put more succinctly via scientific notation, 10^{-33} seconds after the birth
of the universe). The model gets us tantalizingly close to the beginning.
Yet the story is still frustratingly incomplete. After all, the most interesting question that we can ask is not what was happening some tiny fraction
of a second after the big bang, but rather, can we turn the clock all the way
back to "time zero"?

That last step is a big one. How did time begin? What, if anything,
was going on *before* the big bang? Does it even make any sense to ask
that question? Physicists and cosmologists are used to hearing the issue
raised. At the end of countless public lectures on cosmology, someone in
the audience invariably comes up to the microphone, thanks the speaker
for an insightful talk, and then asks, "So, what happened *before* that?" For

many years, the question was simply considered off limits to science. Physicists and cosmologists, it was suggested, could probe what came *after* the big bang but were powerless to probe the actual origin of the universe; the matter was better left to philosophy or religion. Many physicists, meanwhile, argued that the big bang marked the beginning of time, so there simply was no such thing as "time" beforehand. The question of what happened before the big bang then becomes meaningless, just as one cannot meaningfully say what lies north of the north pole.

Our quest to understand that first moment is thwarted on several fronts. To begin with, our theoretical tools are inadequate. Gravity is the force that governs the expansion of the universe, and we understand gravity reasonably well, thanks to Einstein's general relativity. But gravity by itself will not be enough to probe the universe's earliest moments. The equations of general relativity tell us that at the moment of the big bang everything in the universe would have been *infinitely* squeezed together; the universe would have been compressed into a single point. In the jargon of mathematical physics, such a point is known as a "singularity." To a physicist, singularities are about as welcome as the plague. A theory that describes the real world shouldn't have such mathematical kinks, they reason, and so a model that predicts the existence of singularities is treated with suspicion. We will ultimately need not only general relativity but also quantum theory, the theory of the very small. Perhaps when these two ways of describing the world are brought together, in the long-sought theory of quantum gravity, the infinities will disappear, and we will have a coherent picture of the beginning of time.

One suggestion, put forward by Stephen Hawking and U.S. physicist James Hartle, uses quantum theory to blur time and space together, on a microscopic scale, at the moment of the big bang. Known as the "no boundary" proposal, it eliminates the problem of the singularity by "smearing out" time's origin. (Hawking gives an outline of the proposal in Chapter 8 of *A Brief History of Time*, for those who made it that far.) In this model, although time does not stretch indefinitely into the past, it also has no sharp beginning. Time does not suddenly "switch on" but rather emerges gradually from space.

For now, the no-boundary proposal remains just that – a proposal. So far, a full-fledged theory of quantum gravity – at least, one that everyone can agree on – does not exist. Of the various efforts to achieve such a framework, the front runner is *string theory*, an ambitious attempt to unify gravity with the other forces of nature.* In the string picture, the most fundamental "bits" of matter are not particles, but rather tiny loops of string. Because the strings have a finite size, the problem of infinite compression – those pesky singularities – is avoided. And the theory does seem to give a quantum description of gravity. But string theory actually gives physicists more than they asked for: it seems to indicate that we live in a world of more than three dimensions (more accurately, three dimensions of space and one of time). In fact, it says we might live in a world of ten or eleven dimensions.

The idea of extra dimensions sounds bizarre at first, but string theorists take it quite seriously. The latest ideas about the role that such dimensions might play in our universe come from a recent spin-off of string theory known as M-theory (where M stands for "membrane"). According to M-theory, in addition to one-dimensional strings, the universe is also made up of membranes with two or more dimensions.

The Universe on a Brane

We will not examine M-theory in detail, but it is worth mentioning that, in the last few years, physicists have begun to develop cosmological models based on the remarkable framework suggested by the theory. In some of these scenarios, our universe – everything we can see through our largest telescopes – might just be a kind of three-dimensional "slice" of some

---

* A more detailed look at string theory can be found in Chapter 7 of *Universe on a T-Shirt*.

higher-dimensional structure. In the jargon of M-theory, we may live on a "brane" (short for "membrane"). (For a comparison, think about your own shadow: when you look at the ground on a sunny day, your shadow is just a two-dimensional slice, so to speak, of your three-dimensional body.)

These brane-world models offer a startling new description of the cosmos. In some scenarios, the entire visible universe is merely a three-dimensional membrane – called a "3-brane" – embedded in a larger structure, known as the "bulk." This larger structure must have at least four space dimensions – and, as usual, one more for time. The remarkable thing about these brane models is that there's no reason to presume that our universe is unique: there could be any number of "parallel" universes – parallel branes – sitting alongside ours in the four-dimensional bulk. Even Hawking, once a skeptic, now seems to back the brane-world idea. At the conference in Davis, he said, "I must admit I have been reluctant to believe in extra dimensions, but the M-theory network fits together so beautifully, and has so many unexpected correspondences, that I feel to ignore it would be like claiming that God put fossils in the rocks to trick Darwin into believing in evolution."

One of the pioneers of brane-world cosmology is Paul Steinhardt of Princeton University. Steinhardt was one of the first people to work on the inflation theory, but in the last few years he's focused much of his energy on exploring these brane-world models of the cosmos. One particular model that he developed, in collaboration with Neil Turok (now at the Perimeter Institute) and several other colleagues, offers an entirely new view of the big bang. There are two actually two versions of this idea, known as the "ekpyrotic" and "cyclic" models. Both of them describe the big bang as a collision between two of these branes. The big difference is that in the cyclic model, such collisions can happen again and again. The result is multiple big bangs, leading to a potentially endless series of universes.

After a very full day at the Davis conference, I sat down with Steinhardt in his hotel room, where he tried to help me picture these new models of the universe. "You should imagine that there's a force between these two 3-dimensional worlds that would tend to draw them together, as if they were two rubber sheets being drawn together by a spring," he said. "And at regular intervals they would come together, smash together, creating a certain amount of heat – which we would think of as being radiation and matter – and then bounce apart again." In other words, if we lived on one of those branes, we would have the impression that there was a massive burst of energy in our past – which is, in fact, what we perceive. In the cyclic version, he explains, the "bounce" is both an ending and a beginning – the end of one "universe" and the beginning of another, as it enters a new phase of expansion and cooling. "Now we begin the cycle anew," Steinhardt says. "We have the universe filled with hot matter and radiation, we form new stars, new galaxies, new planets, presumably new life – and the cycle continues."

As I tried to wrap my brain around these branes, I reminded myself that Steinhardt's proposal is more educated speculation than testable theory – at least for now. But if the cyclic model is correct, it clearly says something quite profound about both space and time. The theory "raises the basic issue of whether time has a beginning at the moment of big bang," Steinhardt says, "or whether the big bang is instead really a transition to an earlier epoch of evolution."

Steinhardt's brane-world cosmology is not the only attempt to reach beyond the big bang. Guth's inflation theory has evolved, and in one intriguing version, it, too, attempts to describe a larger cosmos. The idea is that the inflation process could have spawned more than just one universe – a model sometimes called "eternal inflation." The most prominent supporter of this view is Andrei Linde, a Russian-born physicist now at Stanford University. Linde believes that whatever caused that period of inflationary growth could have given rise to many – perhaps an infinite number of – separate universes. The cosmos as a whole, Linde speculates, may be immortal. For his part, Guth seems to relish this new, "bigger" version of inflation. And he admits that it would be "clearly an oversimplification to view the big bang as being the beginning of time."

Before the Beginning

These competing visions of our cosmic origins, speculative as they are, must be admired for at least attempting to probe the physics of the early universe beyond the moment of the big bang. They also challenge the imagination as much as anything science has offered. In fact, just about *any* theory of how the universe began necessarily presents such a challenge. We struggle in vain to imagine time stretching back to infinity – but we find it is just as difficult to imagine that time had a beginning.

Part of the problem is that our intuition tells us that every event had a cause; every "happening" is the result of some earlier happening that gave rise to it. And yet quantum mechanics has already shown that some events, such as the decay of a radioactive nucleus, just "happen." Perhaps the universe – along with space and time – did, roughly speaking, pop into existence from nothingness. As physicist Edward Tyron has put it, "Our universe is simply one of those things which happen from time to time." If that picture is correct, then Augustine was on the right track when he declared that God created time and the world together; before that, time simply did not exist.

In contemplating time's origins, our intuition can take us only so far. We're used to dealing with space on the scale of meters and kilometers, and we're used to dealing with time on the scale of seconds and days and years. On these human scales, time seems quite well behaved. Words like "before" and "after" seem perfectly clear in their meaning, while time, like space, seems smooth and uniform – so much so, that we sympathize with Newton and his claim that time "flows uniformly." But Einstein, as we've seen, has already shown us that under certain circumstances, time can behave in much stranger ways. Add to that the shadowy world of quantum mechanics, and we are left with a hint of where physics may be heading.

Physicists hope to one day have a unified theory that reaches beyond both relativity and quantum theory – and when they do, time might not be a part of it. As physicist Lisa Randall has written, we are finding "tantalizing hints of space breaking down at short distances and time breaking down at singularities"; these apparent breakdowns "tell us that fundamentally, space and time are not what we think." Indeed, many physicists now routinely speak of time and space emerging from something – no one can quite say *what* – at the moment of the big bang.

Making Time (Out of Nothing at All)

The idea of "emergent time" is embraced by some of today's brightest young researchers. Among them is Nima Arkani-Hamed, who, at age thirty-six, recently left a tenured position at Harvard to join the faculty of the Institute for Advanced Study in Princeton, New Jersey. "We don't know what happened at the big bang," Arkani-Hamed admits. "But one thing we know for sure is that the whole idea of space and time breaks down – so the idea of what came 'before' may not even make sense." We are "clearly missing something very big," he says – and that something "is certain to involve the idea of emergent time."

Time, in other words, may not be a fundamental part of our universe. Just as the wetness of water emerges from the bulk properties of billions of water molecules sliding past one another, so time itself may emerge from some more fundamental "stuff," whatever that stuff may be.

Exactly *how* time emerges remains to be seen. Time seems to get treated differently depending on whether one starts with general relativity and then tries to add quantum theory, or whether one starts with quantum theory and then tries to add general relativity. (In string theory, which takes the latter approach, physicists usually have to "insert" time and space by hand. They would ideally like to find a "background independent" version of the theory that might actually explain the emergence of space and time based on the vibrating strings or membranes that lie at the heart of the theory.)

David Gross, a particle physicist and string theorist based at the University of California at Santa Barbara, believes the idea of emergent time is something we may simply have to get used to. Gross, who shared a Nobel Prize in 2004 for his work on quarks and the strong nuclear force, believes that the quest for a "unified theory" is leading us toward a picture in which time may no longer be fundamental. "Time is really built in deeply to the way we think about physics, and so is space," he told me recently. "But we've learned in many cases that space is sort of an emergent concept. There are realms of string theory that are best formulated in terms of concepts that don't involve space at all. We have no such examples involving time – but since space and time are so connected in our understanding, it's hard to imagine that if space is an emergent concept, that time wouldn't be as well." This would change, among other things, the way we think about the beginning of the universe. "The idea of time 'flowing' from the beginning until now is an approximation," he says. "It might be a very good approximation to describe the evolution of the universe after a few seconds, but not before."

I press him on the question of what came before the big bang. "There are only three possibilities that I can think of," he says. "One is that, through quantum mechanics, the universe emerged out of nothing. Or else there was something there before. Or else you change the question – which is my favorite," he says. In other words, time "is an emergent concept that doesn't make any sense under those circumstances. So these are the three possibilities that I can see. And, who knows?"

It's only natural for theorists at the frontiers of physics to push ahead with all manner of avant-garde ideas, from string theory to brane-world cosmology and beyond. Yet without grounding in experiment, such musings can only go so far. One of the frustrating things about cosmology, of course, is that one cannot repeat the experiment: the big bang happened once – at least, one time that we can be sure of – and we have to make do. But while scientists can't recreate the big bang, they are hoping to do the next best thing: at a massive new particle accelerator near Geneva, Switzerland, physicists are trying to reproduce the intense temperatures and energies that prevailed in the earliest moments of the universe. The $10-billion project is called the Large Hadron Collider, or LHC, and it was nearing completion just as this book was going to press.

A lot is at stake at the LHC. Experiments at the collider could hint at how the forces that we see today were unified in the remote past. They could reveal exotic properties of matter like "supersymmetry," and may shed some light on the dark matter that makes up much of the universe. We may also get a glimpse of the extra dimensions beyond the three dimensions of space that we're familiar with; if such evidence is found, string theory may not seem like such a stretch of the imagination after all. And then there's the "Higgs boson," the particle believed to be responsible for mass. (The Higgs is thought to generate a field, analogous to an electromagnetic field, which in turn makes other particles seem heavy.) Many physicists are confident that the LHC will finally snare the Higgs, jokingly referred to as the "God particle."

Of course, it will take months – even years – to wade through the data once the machine is up and running, and there are many different kinds of experiments to perform. But most scientists seem confident that the LHC will finally provide answers to some of the deepest problems in physics – perhaps even illuminating the difficult issue of the emergence of space and time. "We'll have something to say by 2010," asserts Arkani-Hamed.

The Arrow of Time Revisited

Physicists struggle not only with the question of how time began, but also with the equally vexing problem of how it came to have a *direction*. We have already looked at the "arrow of time" in connection with entropy and the second law of thermodynamics – but the second law, as we've seen, seems to provide only a partial explanation for time's elusive flow.

The thermodynamic arrow of time points from order to disorder, from low entropy to high entropy, from teacups to shattered china. But time's arrow also reveals itself in other ways. In fact, scientists have noted at least six different (though perhaps related) "arrows of time." Beginning with the one just mentioned, they are:

• *The thermodynamic arrow of time*

The second law of thermodynamics says that the amount of disorder in a closed system must increase over time. Typical examples include the breaking of an egg, the mixing of coffee and cream, or the melting of a block of ice – but the same principle is borne out when we observe any complex process in nature.

• *The radiative arrow of time*

Think of a rock thrown into a pond: the impact creates circular ripples in the pond's surface, and these ripples move outward from the point of impact, in circles of ever-increasing size. We do not see the reverse process: we do not see slight distortions at the edges of ponds slowly moving toward each other, gaining strength and speed, until they converge at a spot in the middle of a pond, causing a rock to be ejected upward from the pond's surface. But the equations work either way: the mathematical description we use to analyze the waves does not indicate a preferred direction for their motion.

It is not just water waves that display this preference. Maxwell's equations describe the propagation of electromagnetic waves – but again, they do not tell us which way the waves will move. (For example, they would be equally valid in describing light waves coming together from deep space to merge at a camper's flashlight, rather than the reverse.) In physics, the normal forward-in-time propagation of waves that we see in nature is said to produce "retarded waves" (because they arrive late, so to speak), while the reverse scenario involves "advanced waves" (which, if they existed, would arrive early). The advanced waves are allowed mathematically but never seem to occur. As with the thermodynamic arrow, probability appears to play a role: the chances of those ripples from the sides of the pond coming together "just so" would be staggeringly unlikely. Indeed, such a wave would cause the entropy of the system to decrease (as would an advanced electromagnetic wave). Because of this connection, some physicists believe the radiative arrow can be explained by means of the thermodynamic arrow.

• *The quantum arrow of time*

As we saw in Chapter 7, quantum mechanics may provide yet another arrow of time: when a quantum system is observed, the wave function of the system is said to "collapse" from a superposition of many states into a single state. This collapse appears to be irreversible, suggesting a link to the direction of time. It is not clear how this arrow may relate to the others, although some have speculated on a link between the quantum arrow and the thermodynamic arrow.

• *The kaon arrow of time*

Just about every subatomic process that we know of is, in principle, reversible; the mathematical descriptions of the behavior of the particles involved indicate that they can "go" either way, with no preferred direction in time. But there seems to be one peculiar exception – a particle known as the "neutral k-meson," or "kaon." (There are also positively charged and negatively charged versions of the particle.) The neutral kaon is unstable; it quickly decays, usually into a similar kind of subatomic particle called the "pion."

The decay process, governed by the "weak" nuclear force, can go both ways; that is, physicists can smash pions together to create kaons. But there is a difference: the reaction used to produce kaons takes just a trillionth of a trillionth of a second (10^{-24} seconds), but the decay takes longer – up to a nanosecond (10^{-9} seconds). Why should the decay of a kaon take a thousand trillion times longer than its creation? (As Paul Davies remarks, it is "rather like throwing a ball in the air and finding it takes a million years to come down again.") The kaon's peculiar inclination to "play by its own rules" is deeply puzzling, and there is no obvious connection between this "kaon arrow" (sometimes known as the "weak interaction arrow") and the other arrows of time. (Even so, the significance of this arrow is disputed. Brian Greene, for example, has said that the behavior of kaons is "likely to be of little relevance" to the arrow of time.)

• *The cosmological arrow of time*

The universe has been expanding ever since the moment of the big bang, about 14 billion years ago. Many physicists would say that this defines a "cosmological arrow of time" – one direction points to our hot, dense past; the other, to our cool, sparse future.

As we've seen, some physicists suspect a link between the cosmological arrow and the thermodynamic arrow, as both seem to be the result of the peculiar conditions that prevailed in the early universe; we will look at this more closely in a moment.

• *The psychological arrow of time*

Finally, there is the arrow of time suggested by our direct experience, based on our perceptions of the world: we remember the past but not the future; we experience – or seem to experience – a "flow" of time in a unique direction. When the brain is thought of as an information-processing system – a set of correlations between the billions of neurons that make up the brain – a link is suggested, perhaps, between the psychological arrow and the thermodynamic arrow. (Hawking, for one, supports this view.)

Physicists (as well as philosophers and psychologists) have spent years grappling with the question of how these seemingly unrelated arrows of time may be connected. Probably no one has considered those connections more deeply than Oxford mathematical physicist Roger Penrose. Described recently by *Discover* magazine as a "polymath extraordinaire," Penrose first made a name for himself through his work on black holes. In the 1960s, in collaboration with Stephen Hawking, he showed that the collapse of a massive star would inevitably lead to a singularity, and that a singularity must be surrounded by an "event horizon" – the zone surrounding a black hole from which nothing can escape. He has also developed a novel description of spacetime known as "twistor theory," which suggests that space and time are "quantized" rather than continuous – that

Physicist Roger Penrose. *(Courtesy of Jerry Bauer)*

is, both space and time are made up of discrete chunks – and that those chunks can be described by means of imaginary numbers (such as the square root of –1).* Penrose has also made contributions to pure mathematics: in the 1970s he proved that one can cover a flat surface with tiles in such a way that the pattern you create never repeats itself, even if your tiles come in just two different shapes; this was previously thought to be impossible. These are now called "Penrose tiles."**

---

* As odd as it may seem to the nonspecialist, physicists often make use of imaginary numbers, which have proven to be particularly useful in solving problems in electromagnetism and quantum theory.

** It turns out that Penrose tiles have practical applications: in physics, they describe a kind of crystal known as a "quasicrystal."

The Singular Scientist

I remember seeing Penrose in action for the first time in the spring of 1990. I was still in journalism school at the time, and Penrose was in town to give a public lecture at the University of Toronto. I vividly recall his use of overhead projection sheets – already somewhat "old-school" at that time. The diagrams and text were done by hand with thick colored marker. There were spacetime diagrams, cartoon-like alive-and-dead cats, and somewhat intimidating rows of equations; each sheet was a little more challenging than the one before. He talked about Gödel and Plato, computers and algorithms, brains and minds. In the lobby after the talk, I bought a copy of his just-released book, *The Emperor's New Mind*. The book – described as "brain-aching" by the *New York Times* – kept me occupied through much of the following summer. His latest book, a 1,100-page tome titled *The Road to Reality: A Complete Guide to the Laws of the Universe* (2004), is even heftier.

Over the years, I've had the chance to speak with Professor Penrose, now in his mid-seventies, on several occasions, most recently on a visit to Oxford in the spring of 2007. Arriving early, I take a stroll through the triangular-shaped park across from the Mathematical Institute. At one end is a small parish church dedicated to St. Giles. Everything in Oxford seems to speak across the centuries, and St. Giles does not disappoint. A plaque on the wall lists all of the vicars going back to 1226. In the park there's a small graveyard. Some of the tombstones are so weathered as to be completely illegible. Beyond the graves, at the tip of the triangle, there's a war memorial; locals sit on the lower tiers of the monument to eat their lunches, talk on their cell phones, and enjoy the May sunshine. Between the graveyard and the memorial there's a spherical bronze sundial – which reminds me it is time to go inside.

Some leading physicists, should they happen to stray off campus, could pass for lawyers or accountants (or in a few rare cases, rock musicians). Not Roger Penrose. Wearing a navy blue sweater and a tweed jacket, he looks like the very definition of "long-tenured theoretical physicist." For a professor, though, his office at the Mathematical Institute is unusually tidy, with books and journals arranged in neat rows and stacks.

On his desk I notice two cardboard boxes filled with the Penrose tiles that he invented thirty years earlier. One set is an amusing variation known as "Penrose chickens." Think of it as a jigsaw puzzle containing two kinds of shapes – fat chickens and thin chickens. Penrose licensed the design to a company that specializes in mathematical puzzles and games, which crafted them using thick, brightly colored sheets of plastic. Penrose periodically picks up a chicken, or casually slides one around on the desk, as we begin to talk about the nature of time.

We start with the arrow – or rather arrows – of time. Penrose assures me that they are real. The flow of time is another matter; that may well be in our heads. In *The Emperor's New Mind*, he writes, "We *seem* to be moving ever forward, from a definite past into an uncertain future … Yet physics, as we know it, tells a different story." And later, in his second popular book, *Shadows of the Mind* (1994): "According to relativity, one has just a 'static' four-dimensional space-time, with no 'flowing' about it. The space-time is just *there* and time 'flows' no more than does space." But the appearance of an arrow of time – that is, a unique "directionality" associated with that apparent flow – is real enough, he says. The various arrows are the result of physical phenomena that we can observe and measure. "Many of them are related to each other – [although] not always terribly simply," he says. Some of them, such as the arrow of time suggested by the decay of the kaon, are "still very puzzling." The psychological arrow, too, is so far unexplained. Even though *Shadows of the Mind* dealt with the problem of consciousness, he admits we do not yet have enough of a grip on the problem to make much headway. "Remembering the past rather than the future – I don't think we understand enough about consciousness to say too much about that," he says.

One area where we have made progress is with the thermodynamic arrow; it is certainly the arrow that Penrose has studied most intensely. As we saw in Chapter 6, however, the origin of the thermodynamic arrow of time presents somewhat of a mystery. The second law of thermodynamics tells us that if we currently have a low-entropy system, we can expect to have a high-entropy system in the future – but it does not tell us where the present low-entropy state came from. Can it ultimately be traced back to the origin of the universe itself? Perhaps the origin of the thermodynamic arrow of time lies in the nature of the big bang.

A Very Special Big Bang

"We really have to go back and look at initial conditions," Penrose says. "The second law of thermodynamics tells us – in very sort of commonplace terms – that things get more random as time goes on." But as you look back to earlier and earlier times, this leads to a problem. "It tells you that as you go back in time, things get less and less random" – that is, they become more and more ordered; the entropy decreases. But, Penrose explains, this seems to be at odds with our observations, which suggest just the opposite. Our clearest picture of the early universe comes from the cosmic microwave background radiation (the CMB, which we looked at in Chapter 9). Data from the WMAP satellite and other observations of the CMB show that the initial fireball was actually incredibly "smooth." In the jargon of physics, it was in a state of "thermal equilibrium" – that is, every part of that microwave glow is at almost exactly the same temperature as every other part. If we accept Penrose's line of reasoning – and not all physicists do – then the early universe must have been in a state of very *high* entropy – just the opposite of what we would have expected based on the second law.

"Thermal equilibrium is the maximum entropy state," Penrose says. "In other words, it's the most completely *random* state you could be looking at. Now, this would seem to be a blatant paradox."

Penrose seems genuinely troubled – to the extent that one can be troubled by events that happened 14 billion years ago. He leans back in his chair, his right elbow on the desk. From time to time he picks up a Penrose tile and flips it about. Then he leans back even farther. "I don't know why people haven't worried about this more," he says.

Penrose thinks he knows where we've been led astray: we have failed to take into account gravity. Only when we incorporate gravity into our picture of the early universe, he argues, will we come to understand the roots of the second law of thermodynamics.

When we talk about entropy in everyday situations, we can safely ignore gravity. Usually, we can readily see what's in equilibrium and what isn't. (Think of milk thoroughly mixed in a cup of coffee; the molecules simply can't *be* any more random, so we're confident in saying it's in equilibrium.) At first glance, a perfectly smooth entity like the CMB would also seem to be in equilibrium. But gravity changes things. Although the reasons are rather technical, once we take gravity into account, a perfectly smooth entity like the CMB might actually be seen as being very far from equilibrium – and thus very *low* in entropy.

Penrose concludes that the gravitational field of the early universe was not in equilibrium at all; in fact, he says, "it was very, very special" – and by "special" he means it was in a highly ordered state. How special? At this point, Penrose takes the conversation in what seems like a new direction: he begins to discuss the entropy of black holes. It may sound as though he has gone off topic, but he has not: the big bang and black holes are in some ways very similar, at least mathematically. In one case, matter emerges from a singularity; in the other, matter evolves *into* a singularity. (However, they are not quite mirror images of each other: Penrose cautions that "the singularities in black holes don't look the remotest bit like the big bang in reverse time." I decide to take his word on this.) Fortunately, we do know something about the entropy of black holes – Stephen Hawking and physicist Jacob Bekenstein showed how to work that out in the 1970s – and we can use the same approach for the early universe, Penrose says.

"We can now estimate how special that initial state was," he says. "And it's fantastically special – very, very extraordinarily special. If you just take into account the part of the universe we see, the probability of that initial state having arisen purely by chance is less than 1 part in 10 to the power 10 to the power 123" – that is, 1 part in $10^{10^{123}}$. (This number is so large – a one followed by 10^{123} zeros – that we could not write it out even if we attached a zero to every atom in the known universe.) "Now, that's ridiculously special," Penrose says. "So, there is a huge puzzle there, which cosmologists hardly ever address – and I find this very strange." Even the theory of cosmic inflation, which explains several other problems associated with the physics of the early universe, "does not explain that initial very special state," he says.

Is there any way out of this conundrum? Ultimately, Penrose says, any attempt to develop a unified theory of physics will have to take all of this into account. Indeed, such a theory may have to have some kind of time-asymmetry built into it, he says – which would make it radically different from the laws of physics that we have developed so far. ("I've been saying this for decades," Penrose says, looking more detached than weary. "People listen to me, maybe even nod their heads – then they go off and do the same quantum gravity they've been doing before. They barely pay any attention to this.")

Of course, Penrose is the first to admit that his own approach to quantum gravity is not particularly orthodox: he feels that quantum theory as it currently stands is incomplete, so just trying to "quantize" gravity (Einstein's general relativity) will not work. Penrose does not have the answer, though he would obviously be pleased if twistor theory, or something like it, proved to be on the right track. But he does have his suspicions: he wouldn't be surprised if the time-asymmetry of the "collapse of the wave function" in quantum mechanics turns out to be connected to the time-asymmetry inherent in the second law of thermodynamics – that is, he suspects a link between the quantum arrow and the thermodynamic arrow. "I also believe – and this is a little bit more far out – but I also believe it's got something to do with our perception of the passage of time."

Mind and Matter

Penrose has come up with some unorthodox ideas over the years, but he is always very conscious of what constitutes mainstream science and what does not. He readily admits when one of his ideas rests more on conjecture than on established theory. Now that we are beginning to wade into the prickly question of human consciousness, he knows we are crossing into the realm of speculation. "Does consciousness have anything to do with this? This is where I'm a little more far out on the fringe, to say that, yes, it probably has got something to do with it."

He wonders in particular about the connection – if there is one – between the thermodynamic arrow of time and the psychological arrow. Memory, of course, seems to be intimately connected to time. Yet it operates in just one direction: we remember the past but can only imagine the future. The second law of thermodynamics, meanwhile, seems to work the other way around. If you see an ice cube sitting on the kitchen counter, you know that in a few minutes there will just be a puddle of water. But if someone comes in a bit later and sees only the puddle, they would have no way of knowing where the water had come from. You can't "retrodict" the presence of the ice cube.

"Retrodiction is terrible – normally," says Penrose. "That's what the second law is telling you. Yet remembering is *what we do*. We retrodict with our minds all the time." Ultimately, he says, thermodynamics can offer only limited insight into our mental conception of time's passage. "There isn't a clear statement to say, 'Okay, it's the second law of thermodynamics.' That's no answer. There's something much more subtle going on . . . It has to do with awareness, it has to do with consciousness. It has to do with issues we're very far from understanding."

As for time itself, Penrose stops short of offering a definition. "I really don't know," he says. "I certainly would say that time is not what we think it is. It's not a sort of steady progression – certainly not a sort of universal steady progression."

It is telling that physicists find it easier to say what time isn't than to say what time is. It isn't merely a steady flow, it isn't just an increase in disorder, it isn't simply the reflection of an expanding universe. After speaking to so many scientists, it's beginning to seem as though the only universally-agreed-upon statement one can make about time is that it's not what we think it is.

Even when we do seem to have some particular aspect of time nailed down – such as the "arrow of time" identified by Eddington more than eighty years ago – it threatens to slip from our grasp. Thermo-dynamics seems to illuminate one facet – but only one facet – of time's arrow. We now know that there are many "arrows of time," and even today's brightest minds cannot say precisely how they are related. Perhaps time is less like the singular flow of the Nile and more like the multiply-branching Amazon – or the chaos of a Los Angeles freeway interchange. Or maybe no metaphor is quite powerful enough to capture time's essence.

ALL THINGS MUST PASS

The ultimate fate of life, the universe, and everything

Perceivest not
How stones are also conquered by Time?
Not how lofty towers ruin down,
And boulders crumble? Not how shrines of gods
And idols crack outworn?

— Lucretius, *On the Nature of Things* (1st century B.C.)

Eternity is very long, especially toward the end.

— Woody Allen

The city of Oxford is known for many things – its "dreaming spires," as a poet once described it; its storied quadrangles, honey-colored walls, and graceful arches; its throngs of tourists. The well-preserved college buildings are beautiful and ornate – and of course old. So old, the guidebooks remind us, that names like "New College" can be misleading: the college was founded in 1379, when the university itself was already a couple of centuries old. There is a wonderful story about the wooden roof of the college's dining hall. In the middle of the nineteenth century, the massive oak beams that support the roof had to be replaced. The warden and his team of carpenters promptly went out to the forests owned by the college and cut down mighty oak trees that had been planted five hundred years earlier – just a few years after the founding of the college – for that very purpose.

The story may be slightly exaggerated – apparently oaks are typically harvested after 150 years, not 500 – but it is still an inspirational story of long-term planning of the sort rarely seen in the frantic bustle of today's world. One of those touched by the story was Danny Hillis, an American inventor and scientist known for his pioneering work with supercomputers. Hillis began to reflect on humanity's short-sightedness, and considered different ways to encourage long-term thinking. Eventually he decided on a machine: he vowed to build a clock that would run for 10,000 years. Hillis, who once ran the research and development department of Walt Disney Imagineering, teamed up with a group of like-minded thinkers who shared his vision of "deep time." The group came to include Kevin Kelly, the founding editor of *Wired* magazine, and Stewart Brand, creator of the Whole Earth Catalog, who once said that "civilization is revving itself into a pathologically short attention span." The group established the Long Now Foundation, and their planned clock has been dubbed the Clock of the Long Now. (The name comes from British avant-garde musician Brian Eno, who coined the phrase "long now" after a visit to New York. The right-here, right-now perspective of New Yorkers struck him as radically different from the longer view taken by Europeans. "I came to think of this as 'The Short Now,' and this suggested the possibility of its opposite – 'The Long Now,'" Eno has written. "'Now' is never just a moment. The Long Now is the recognition that the precise moment you're in grows out of the past and is a seed for the future.")

Embracing the Long Now

The long-term plan is to build this great chronometer in the Nevada desert, where Hillis's foundation has already bought a seventy-five hectare plot. The clock, they imagine, will be a towering structure, perhaps twenty-five meters in height, and will be built to endure through the ages. That project is still in the planning stages, but a more modest prototype, just three meters high, has already been built. The device was assembled at the group's workshop in northern California, and now rests in the Time Gallery in London's Museum of Science. It is an appropriate home. On the gallery's upper level is a staggering array of clocks and watches from across

The prototype of the Clock of the Long Now. The clock is designed to keep time for 10,000 years — note the five-digit display for the year. *(Photos by author)*

the ages, from a replica of an Egyptian shadow clock dating to around the ninth century B.C. to the first cesium atomic clock, built in 1955. But the Long Now clock does not resemble any of these.

"It doesn't really look like what you would expect a clock to look like," says Alison Boyle, the museum's curator of astronomy and modern physics. As we approach the clock's glass-walled display case, it looks like some kind of top-heavy, one-eyed metal creature. The giant black "eye," about half a meter across, is the clock's dial, which displays just about everything except for hours and minutes. ("Hours are an arbitrary artifact of our culture," Hillis says.) Instead it shows a rotating star field designed to reflect the motions of the real sky, displaying the phase of the moon and the position of the sun, even taking into account the 26,000-year precession cycle of the equinoxes.* An outer dial displays the Gregorian year, shown as a five-digit number; 2007, for example, is shown as 02007. "The idea is, obviously, that it will be going for 10,000 years, so we need to avoid a 'Y10K' problem," says Boyle.

Even in the subdued light of the museum, the device gleams; its various parts are made of brass, stainless steel, tungsten, and a nickel-copper alloy that goes by the trade name of Monel. Yet in some ways, the clock is very old-school: there are no electronics; the power comes from falling weights, just as it did in the first mechanical clocks that started to appear in England's cathedrals in the thirteenth and fourteenth centuries. The clock "ticks" twice a day, the motion of its gears regulated by a device known as a "torsional pendulum." The three-pronged pendulum, supporting three massive tungsten spheres, rotates back and forth in a horizontal plane at the bottom of the machine. Above the pendulum is a complicated-looking device known as a "binary mechanical computer" (also called a "serial bit adder") that controls the display on the dial; its shape is suggestive of records stacked in a jukebox, or perhaps a pile of metallic pancakes suitable for an android's breakfast. Of all the artifacts in the museum, Boyle says, the thing it most closely resembles

---

* We looked at precession briefly in Chapter 1 (see page 13).

are the designs of nineteenth-century computer pioneer Charles Babbage; indeed, the Long Now clock has been dubbed "the world's slowest computer." Actually, it is hard to pin down just what era the Long Now machine belongs to: one imagines it was put together by Leonardo da Vinci, using spare parts from the *Millennium Falcon*.

From its desert home, the full-scale version of the clock will read the sun's passage across the sky, automatically remaining in sync with respect to solar time over the millennia. In this respect it is again a throwback to designs of past centuries, when clocks would be corrected, manually, to match the readings of a sundial. Yet the Long Now clock would not be self-sustaining: it would need to be "wound" – a deliberate attempt to turn its caretakers into active participants rather than passive observers. One of the aims is to foster a sense of stewardship and responsibility. Indeed, its parts may need to be periodically replaced.

Already the London prototype has "become a symbol for a whole new way of thinking about time," says Boyle. "The idea behind this clock is not only that it keeps time, but it actually encourages us to change the way that we think about time, and how we measure it."

The prototype was completed late in 1999, and "ticked" for the first time at midnight on December 31 of that year, ushering in the new millennium.* It then started the third millennium with two rings of its one-thousand-year chime. Members of the Long Now team imagine the full-scale version accompanied by a library, with a digital archive of texts in a thousand languages – another attempt to communicate across the ages.

One can think of the Clock of the Long Now as a temporal analog of the brass plaques that accompanied the *Pioneer* spacecraft that traversed the outer solar system in the 1970s – the first purpose-built human artifacts destined to leave the solar system. The plaques, which depict a naked man

---

* At least for those who consider the millennium to have begun at the start of 2000 rather than 2001.

and woman, data about the solar system, and music from Chuck Berry and others, were designed to communicate across vast distances; should they one day be intercepted by intelligent aliens, the plaques would tell them quite a bit about the civilization that sent them (including, if they had the right sort of record player, what we liked to dance to). The Clock of the Long Now is designed to communicate across time. The challenge is similar – very similar, in fact, because the *Pioneer* craft are moving so slowly that by the time they are discovered (in itself a long shot), millions of years will likely have passed. A ten-thousand-year clock forces us to ask what future civilizations will be like. Will human beings millennia from now resemble us in any recognizable way? Will they think and act the way we do? Will they value what we value?

Peering into the future is like straining to see through a thick fog; nearby objects can be seen, at least in rough outline; more distant landscapes are lost in the mist. Time obscures the view.

In the physical realm, the next few moments may seem clear enough – a falling hammer will hit the ground, a rainstorm will eventually peter out. But as soon as we try to make specific predictions about complex systems – and that would certainly include human affairs – we see how limited our powers of foresight are. We know the future will bring both death and taxes, but we cannot say who will die when (probably for the better), or precisely how much tax we will pay a few years down the road.

We can be more confident in the broad brushstrokes than in the details. Population forecasters, for example, can look at fertility rates and predict that the world's population will peak at about 9 billion by 2070 before beginning to fall. We can be certain that most of that growth will take place in developing countries, and that more and more of the world's population will live in cities (more than 60 per cent by 2025, according to UN projections).

Trying to imagine specific events is far tougher. We may be reasonably certain that 2018 will see both a World Series and an Academy Awards ceremony – but what about 2518? There will presumably be a U.S. federal election in 2040 – but what about the year 3000, or 10000? The weekly newsmagazines routinely run cover stories like "Your Life in 2020," looking a decade or so into the future – but who, other than science-fiction writers, really gives any thought to what civilization might be like a thousand years from now? Or a million? The farther ahead we look, the murkier it all becomes. While the past seems etched in stone, the future is a blur of infinite possibility. If the Clock of the Long Now survives ten thousand years – roughly the length of time from the Agricultural Revolution until now – who will gaze upon it?

A Brief History of the Future

Trying to forecast the future is a relatively new pastime. There were, of course, religious texts that made such predictions – although works like the Book of Daniel in the Hebrew Bible and the Book of Revelation in the New Testament also served to deliver more urgent messages about the eras in which they were written. In the sixteenth and seventeenth centuries, the first utopian worlds were imagined – notably in Thomas More's *Utopia* and Francis Bacon's *New Atlantis*. The sixteenth-century French apothecary Nostradamus published collections of prophecies, predicting various kinds of natural disasters along with wars and military invasions – all of them vaguely worded, undated, and open to endless interpretation. Also in France, some two hundred years later, a collection of long-term predictions appeared in a book called *L'An 2440*, published in 1770 by dramatist Louis-Sébastien Mercier. The book tells the story of an eighteenth-century Frenchman who falls into a deep sleep and awakens seven centuries later. He discovers that in the twenty-fifth century, war has been nearly eliminated and slavery has been abolished. France is still a monarchy, but its population has grown by half, and Paris has been rebuilt on a scientific plan. A canal has been built through the Suez, and balloons (!) offer rapid transport from one continent to the next.

Predictions of this sort become more common in the nineteenth century. (By the end of that century, the genre we now call science fiction was beginning to flourish.) One example, again from France, comes from a series of colorful cigarette cards printed in 1900. Commissioned as part of the *fin de siècle* festivities held throughout the country, the cards depict life "*en l'an 2000*." (The illustrations were reprinted in Isaac Asimov's *Futuredays* in 1986.) The whimsical drawings depict all manner of airships and lightweight wooden flying machines – all rather flimsy-looking to the modern eye. Personal flying devices seem to be ubiquitous, though they appear to be little more than canvas or fabric wings that attach to the body; presumably the wings flap to provide lifting power.

It seems as though the artist has taken the technology of 1900 and attempted to extrapolate it into the future – and yet has missed nearly all of the great technological developments of the century that lay ahead. Aviation did, of course, blossom – but thanks to the jet engine and light-weight aluminum alloys, our planes typically carry many people (often hundreds), and do so primarily over long distances. Personal flying devices have not been realized, and most short-distance travel is accomplished by automobile at ground level. The artist also clearly imagined a progression of bigger and better airships – and yet, even before the fiery explosion that destroyed the *Hindenburg* in 1937, the airship was on the way out.

There were many who believed that heavier-than-air flying machines would never succeed. In 1895, Lord Kelvin – at that time, President of the Royal Society – called such craft "impossible"; a few years later, the Canadian-born astronomer and mathematician Simon Newcomb declared, "Flight by machines heavier than air is unpractical [*sic*] and insignificant, if not utterly impossible." The Wright brothers took flight at Kitty Hawk just eighteen months later.

In the twentieth century, the most famous bad predictions seem to involve the computer. In the 1940s, the chairman of IBM suggested that the worldwide market for computers could just about be counted on one hand. And in 1977, Ken Olson, the president of Digital Equipment Corp., said, "There is no reason anyone would want a computer in their home." (The Commodore PET, the Tandy Corporation's TRS 80, and the Apple II were all introduced that very year, and, for better or for worse, people have been hunched over their home computers ever since.) Just a few years later, in 1981, Bill Gates is said to have declared that "640,000 bytes of memory ought to be enough for anybody."

One could go on endlessly with examples of "bad predictions"; the Internet is full of them. Predicting the future – and the future of technology in particular – is a multifaceted challenge. Often the importance of new materials (plastics, aluminum, steel) is not foreseen. Sometimes a new technology – even if it builds on advances that have come before – can seem utterly novel (as with the Internet). Even when a discovery or invention is already upon us, there can be a kind of ripple effect of unimagined consequences. When the first Model T rolled off the Ford assembly line in 1908, who could have foreseen freeways, traffic jams, suburban sprawl, the rise of the shopping mall (and the decline of "Main Street"), grievous air pollution, or global warming? Sometimes change happens faster than we could have imagined; sometimes much slower – as with those flying cars and robotic servants that always seem to be just a decade away, and yet never quite materialize.

A Word from the Futurists

It will be interesting to see if today's generation of futurists fare any better. Science-fiction author Arthur C. Clarke, who died in early 2008, has a mixed record. In a 1971 short story, he placed the first manned mission to Mars in 1994; by 1999 he was more cautious, admitting that "we'll be lucky to make it by 2010." With just two years to go, it seems unlikely we will make it by then, either. (The consensus these days is more like 2050 to 2080.) But Clarke also had some major successes: he predicted the use of communications satellites in geosynchronous orbits, as well as the so-called millennium bug (which, happily, turned out to be fairly benign when the computer clocks rolled over on January 1, 2000).

Some of Clarke's more recent predictions were relatively modest – Prince Harry will become the first member of the royal family to fly in space – while others were more exotic. He expected that, by the end of the century, we'll have a new kind of propulsion system for rockets (a "space drive"), and that human explorers will use it to head for nearby star systems – worlds which, by that time, we will have already explored robotically. At that point, he said, "history will truly begin." Clarke also predicted that artificial intelligence will reach human levels by 2020, after which there will be "two intelligent species on Planet Earth," one evolving much more rapidly than the other.

Stephen Hawking seems to agree that we should be cautious in the face of accelerating computer technology. In an interview in 2001, Hawking said that human beings should change their DNA through genetic manipulation in order to keep ahead of our electronic rivals and stop intelligent machines from gaining the upper hand. "The danger is real," he said, "that this [computer] intelligence will develop and take over the world."

Inventor and futurist Ray Kurzweil agrees that a human-computer "merger" is inevitable. He predicts that by 2019, $1,000 worth of computing power will have capabilities similar to those of a human brain; by 2029, machines will claim to be conscious; and by 2099 there will no longer be "any clear distinction between humans and computers."

For a more optimistic outlook, we can turn to U.S. physicist and science popularizer Michio Kaku. In his 1997 book *Visions*, he presents a detailed look at how science and technology will transform society over the next hundred years. Thanks to advances in physics, biomedicine, and computer technology, he predicts we "are on the cusp of an epoch-making transformation, from being *passive observers of Nature to being active choreographers of Nature*" [Kaku's emphasis]. Many cancers will be curable by 2020, he says; computers will be as cheap as beer; a day trip into Earth orbit will cost about the same as a transatlantic flight. (With twelve years to go, however, some of these may be a close call.) It's a rosy picture overall, though as Kaku acknowledges, "In the background always lurks the possibility of a nuclear war, the outbreak of a deadly disease, or a collapse of the environment." Well, those *would* put a damper on things, wouldn't they?

The End of Civilization

Ah, yes – the collapse of civilization: a perennial favorite for scholarly analysis and speculation of all kinds. As we contemplate the vast stretches of time that lie ahead, it's natural to wonder if human beings will be a part of that future. Will our civilization march forward, or come crashing down? Worrying about the end of the world has been a popular pastime for nearly as long as there have been humans on our planet. Ancient cultures had no end of stories of apocalyptic destruction, and one could say that the Scientific Revolution merely put a new spin on such fears: instead of waiting for the gods to wipe us out, it raised the possibility that we would simply do the job ourselves. Scientific advances have, of course, lengthened our lives and increased our numbers; but they have also shown us, for the first time, pathways that could lead to total annihilation. Ecologist Doug Cocks has described our condition as "an escalating struggle between knowledge and catastrophe." From the Bible to the bomb to global climate change, we've always found ways to imagine our downfall.

The last few decades have brought new kinds of anxieties. In his provocative book *Our Final Hour* (2003), British physicist Martin Rees outlines some of his most urgent concerns. Until now, he says, only a nation – or at least an angry province or rebellious group – had the power to unleash havoc on a large scale. Now, thanks to advances in technology (especially biotechnology), Rees says we're entering an era in which "a few adherents of a death-seeking cult, or even a single, embittered individual, could unleash an attack."

As Rees reminds us, it's not just militant fundamentalist groups like al-Qaeda that we have to worry about; we should also fear smaller but equally deadly cults like Heaven's Gate (responsible for a mass suicide in 1997) and Aum Shinrikyo (the group behind the 1995 Sarin gas attack on the Tokyo subway), as well as angry individuals like the Oklahoma City bombers and the Unabomber. A single person, he stresses, can now become a deadly force even without a cult of like-minded followers. "There will always be disaffected loners in every country," Rees warns, "and the 'leverage' that each can exert is increasing."

Another danger, Rees says, is that society is becoming increasingly integrated and interdependent. No disaster can be truly "local" anymore: what affects one city, state, or province will automatically affect attitudes and behavior around the world. Nothing illustrates this better than the SARS outbreak of 2003: an infection that began in Asia quickly found its way to Canada's largest city, and TV images of a few people wearing masks in Toronto instantly tainted the city's reputation thousands of miles away. It took several years for the local tourism industry to recover.

But all of these dangers, Rees says, may be temporary – at least from a long-term perspective. Soon – perhaps before the end of this century – human civilization will spread beyond Earth. Once that happens, it is very unlikely that any one disaster, no matter how severe, will destroy our species completely. He argues that we are at a sort of bottleneck in time: today's dangers are substantial and varied – but if we can make it through the next few decades, we may be in the clear for good.

Thinking of Copernicus at the Berlin Wall

A more abstract but equally intriguing approach to humanity's long-term prospects comes from physicist J. Richard Gott of Princeton University (we met him briefly in Chapter 8 when we looked at cosmic strings as a proposed method of time travel). Gott uses what he calls the "Copernican Principle" to predict the longevity of our species – and, for that matter, the longevity of just about anything. The principle is named after Copernicus because the great astronomer showed us that we're not in a "special" place – the earth is just one of many planets, and, as we later discovered, the sun is an ordinary star. Gott asserts that, similarly, we are not in a special *time*. More specifically, if you happen to come across some entity – it really doesn't matter what the entity is – you can safely make two assumptions: you are probably not encountering it right after it came into existence, and you are probably not seeing it just before its demise. (Either of these cases would be "special" times, and therefore deemed unlikely by the principle.) It's much more likely, he argues, that you came upon the entity at some random time in the "middle period" of its existence.

Gott came up with the idea when visiting the Berlin Wall in 1969. The wall had been up for eight years, and many people wondered how long it would last. The Copernican Principle suggests that the best predictor of how long something will endure is *how long it has already lasted*. Gott's reasoning is astonishingly straightforward: he assumed that there was a 50 per cent chance that he was observing the Berlin Wall during the middle half of its existence – that is, between the 25 per cent mark and the 75 per cent mark on the timeline of the wall's history. Doing a little math, he concluded there was a 50 per cent chance that the future longevity of the wall was between one-third as long as its longevity thus far and three times that length.* Since the wall was 8 years old at that time, that meant an interval stretching from 2.7 years to 24 years into the future. As Gott emphasizes, he made no prediction of *how* the wall would come to an end, only of *when* it was likely to happen. When the Berlin Wall fell in 1989 – twenty years after Gott's visit, in accord with his prediction – he decided to write it up. His article "Implications of the Copernican Principle for Our Future Prospects" was published in the journal *Nature* in 1993.

Gott then turned his attention to the human race. He also hiked up his "confidence limits" from 50 per cent up to 95 per cent, the standard traditionally used by scientists. The technique remains the same: under the Copernican principle, there is a 95 per cent chance that you're observing some entity in the "middle 95 per cent" of its existence – that is, between the 2.5 per cent mark and the 97.5 per cent mark on the timeline

---

* Suppose the entity has been around for x years. If you are seeing it at the 25 per cent mark on the timeline of its existence, then its future duration (beginning now) is three times greater than its past duration, i.e., $3x$. If you're already at the 75 per cent mark, however, then its future is only *one-third* as long as its past, i.e., $\frac{1}{3}x$. So your 50 per cent confidence zone (75 per cent–25 per cent = 50 per cent) stretches from a future $\frac{1}{3}x$ in length to a future $3x$ in length. (A parallel argument can be used for the 95 per cent confidence version of the argument.)

of the object's history. Again doing a little math, you can have a 95 per cent confidence that the thing you're observing will last for an interval between 1/39 of its current age to 39 times its current age. *Homo sapiens* have been around for about 200,000 years, so the implication is that we will be around for at least another 5,100 years but probably not more than 7.8 million years. (Those figures, Gott suggests, are in line with the life-times of other hominids: *Homo erectus* lasted for about 1.6 million years, and the Neanderthals for about 100,000 years; mammal species on average seem to last for about 2 million years.)

It all sounds a bit abstract, but Gott has also used the method to predict something more down to earth (and perhaps equally difficult to forecast): the length of the runs of New York theatrical productions. In 1993, Gott predicted the closing dates for forty-four Broadway and off-Broadway plays that were running at the time, based only on their opening dates. Gott tells me that when he last checked the list, forty of the forty-four plays had closed (including *Cats*, which Gott reminds me "was supposed to be for 'now and forever'"), and none of the dates had fallen outside the limits given by the 95 per cent version of the Copernican Principle.

Not everyone has been swayed by Gott's argument. Physicist Freeman Dyson, of the Institute for Advanced Study, says one must be cautious in using "an abstract mathematical model to describe the real world." In particular, if we know that some improbable event has already happened, "then the probabilities of all related events may be drastically altered." Looking at humanity's long-term future, he then raises a point that we heard earlier from Martin Rees: that we may, in fact, live at a *very* special time – that is, humanity may be living just prior to the time when interplanetary travel becomes commonplace. Whether we take advan-tage of this opportunity, he explains, is beside the point. Just knowing that the "rules may change" within the next couple of centuries hampers any use of the Copernican Principle. "The knowledge of this improbable fact changes all the *a priori* probabilities," Dyson writes, "because the escape of life from a planet changes the rules of the game life has to play."

Place Your Bets

You don't need a Ph.D. to play at the Doomsday game. As the twentieth century drew to a close, Hollywood movies about Earth-destroying asteroids filled the screens, and articles on apocalyptic thinking (both religious and secular) became a fixture in highbrow magazines. (Cosmic collisions are particularly frightening because we know they have happened several times before. There's compelling evidence, for example, that an asteroid or comet smashed into the Yucatan peninsula in Mexico about 65 million years ago. The impact is thought to have triggered a drastic change in climate, killing off the dinosaurs and hundreds of other species.)

On the eve of the Millennium, the British bookmakers William Hill started taking bets on how the world will end, with odds worked out for about a dozen popular scenarios. (It's not quite clear how one would collect on such a bet.) The favorite is "war," with odds of 1,000 to 1. Climate change is a longer shot, at 250,000 to 1. Even lower down the list is an alien invasion, at 500,000 to 1. Meanwhile, the Long Now Foundation – the people behind the 10,000-year clock – have set up a website for similar long-term bets (www.longbets.org). There, users can ponder such predictions as "By 2030, commercial passengers will routinely fly in pilotless planes," or "At least one human alive in the year 2000 will still be alive in 2150." (The bets are given as "even odds," with all proceeds going to charity.) Perhaps they will set up a bet over a recent prediction by artificial intelligence expert David Levy: in his book *Love + Sex with Robots* (2007), Levy predicts that by 2050, "love with robots will be as normal as love with other humans, while the number of sexual acts and lovemaking positions commonly practised between humans will be extended, as robots teach more than is in all the world's sex manuals combined."

Predicting how human society will evolve is admittedly fraught with difficulty. Is it any easier when we look at purely physical systems? We saw in Chapter 6 how Newton's laws let us predict how an object responding to a force will move, and how well such predictions are borne out in our solar system. Laplace, as we saw, imagined that if we knew exactly what a physical system was doing – the precise motion of all its component particles – we could establish its future with certainty. When we do have a clear grip on those motions, we can indeed make such predictions: the sun will rise tomorrow; there will be a solar eclipse on August 21, 2017; and so on.

But nature thwarts our efforts on two fronts: quantum theory, as we've seen, prevents us from having perfect knowledge of the speed of any individual particle, let alone a complex system. Secondly, complex systems often evolve in a way that's incredibly sensitive to their initial conditions. (Think again of the "break" in a game of billiards: In the history of the game, have any two breaks ever been identical?) Rewind the system, change the position or speed of some object very slightly, and the subsequent evolution will be different. Such systems are said to be "chaotically unstable." The most famous example is the proverbial butterfly that flaps its wings in the Amazon rainforest, affecting the weather in China months later. (It is quite possible that weather forecasters may *never* be able to accurately predict any particular city's weather more than about a week in advance.) A parallel claim has been made for biological evolution: Stephen Jay Gould argued that if you were to "replay the tape" of the last billion years of life on earth, it would be staggeringly unlikely to see the same creatures (including *Homo sapiens*) emerge in just the same way. Can we be any more confident when we consider the future of the earth itself?

Earth's Final Curtain

Our planet's fate is inexorably tied to that of our sun. Astronomers have understood stellar physics well enough for several decades now to predict, with fairly high confidence, the fate of our home star. It has been shining for some 5 billion years, and may shine for another 5 billion or perhaps a little more. As it uses up its nuclear fuel, however, it will undergo some peculiar contortions. Gravity will at first cause it to shrink in size – but this will make the sun's core hotter, which will actually cause its outer layers to expand significantly. At this stage, the sun will become a "red giant." After a few hundred million years (a short period in terms of the sun's lifetime), it will undergo yet another phase of heating and expansion, shedding much of the material in its outer layers, and finally collapsing into a so-called white dwarf. By this time, its mass will still be about three-quarters of its current value – but compressed into a sphere the size of the earth.

No one has investigated the long-term fate of the earth – and indeed the universe – in more detail than astronomer Fred Adams of the University of Michigan. Adams and co-author Gregory Laughlin described our deep cosmic future in their 1999 book *The Five Ages of the Universe*. When I spoke with Adams in his office in Ann Arbor, shortly before the publication of *Five Ages*, he talked me through our solar system's frightening future. The sun's initial swelling during its red giant phase, he says, will spell disaster. In 5 billion years, give or take, the earth "will no longer be a hospitable place." At that time, the sun – looming in the sky as a giant scarlet orb – will "completely fry the earth." The sun's radius will swell from its current 1.4 million kilometers up to a bloated 168 million kilometers. Given that the radius of the earth's orbit is only 150 million kilometers, this sounds dire indeed. Due to the sun's weakening gravitational attraction, however, the earth's orbit will have expanded by then – out to about 185 million kilometers. So we will not be engulfed – yet – by the swelling sun. What is left of our planet, however, will be scorched beyond recognition, as the sun's luminosity soars to nearly 3,000 times its current level. But we will be in trouble long before then. "Before that happens – in two or three billion years' time – the sun will get hot enough that a runaway greenhouse effect will begin to make life very, very hot," Adams says. Not merely hot enough to make Al Gore downright cranky, but hot enough to boil the oceans. "So in the couple-of-billion-year time scale, the earth itself is in deep trouble as far as life is concerned." In a recent paper, Adams puts it more dryly: "Current estimates indicate that our biosphere will be essentially sterilized in about 3.5 billion years, so this future time marks the end of life on Earth."

The planet itself – now devoid of life – may linger a bit longer. Although Earth will have moved to a wider orbit, Adams explains that it will also meet with more resistance as it passes through "stellar outflow." This will ultimately cause Earth's orbit to decay, dragging the planet closer to the sun, where it will meet its doom. In that same paper, Adams describes the end of Planet Earth in two terrifyingly concise sentences: "Earth is thus evaporated, with its legacy being a small addition to the heavy element supply of the solar photosphere. This point in future history, approximately 7 billion years from now, marks the end of our planet."

Happily, this billion-year time scale is inconceivably long compared to the 200,000 years or so that our species has been around, let alone the few millennia in which we've been using technology. So perhaps we can dare to imagine that we will have spread out across the galaxy, or at least beyond our doomed solar system, before our planet's demise. Let us turn, then, to the long-term prospects for the universe itself.

The Fate of the Universe

If we lived in Newton's cosmos of absolute space and time, it would be reasonable enough to imagine an infinite future either for our species or for our remote descendants, with time ticking away without end. But the discoveries of twentieth-century physics changed that picture. After the big bang model began to solidify in mid-century, astronomy textbooks typically described two possible fates for our universe. If the average density of the universe were great enough, the universe would be "closed": gravity would eventually halt the expansion, and the universe would begin to contract, ultimately collapsing in a kind of reverse big bang known as the "big crunch." If the density were lower than this threshold, the universe would be "open": it would expand forever, and all processes in the universe would gradually "run down" in accordance with the second law of thermodynamics. The universe would become ever darker, colder, and less hospitable to life. A well-known poem by Robert Frost captures the essence of the two possibilities: "Some say the world will end in fire / Some say in ice." Until the final decades of the twentieth century, this was the best we could do: the universe would suffer one of these fates, though we could not say which one. But the universe had more surprises in store, and as the century drew to a close, it delivered a whopper.

In the late 1990s, astronomers were studying the properties of distant galaxies, very much as Hubble had done seventy years earlier (and now using, among other instruments, a space telescope named after him). This time, they focused on exploding stars, or supernovae, within those galaxies, using them to accurately measure their distances. Two international teams, working independently, carried out the surveys. One of them, the High-Z Supernova Team, was led by Brian Schmidt of Australian National University and Adam Riess of the Space Telescope Science Institute in Baltimore. The other team, known as the Supernova Cosmology Project, was led by Saul Perlmutter of the Lawrence Berkeley Laboratory in California. Both teams compared what nearby galaxies were doing with the motion seen in more distant ones. What they found came as a complete surprise. The universe wasn't just expanding – it was accelerating.

The universe appears to have been decelerating until sometime around 7 billion years ago; after that, the universe entered a new era of ever-faster expansion. What could possibly be causing the universe to accelerate? The big bang explosion would have given everything an outward push – but the force of gravity ought to be slowing that expansion; the universe should be slowing down. Astronomers and physicists concluded that there must be some kind of energy that works against gravity – some force that literally pushes all of the galaxies away from each other. No one knows exactly what that entity is; for now it has been labeled "dark energy." It is certainly possible that this dark energy is the energy associated with empty space that Einstein had suggested back in 1917, when he introduced his "cosmological constant." If that turns out to be the case, then his "greatest blunder" was in fact a move of incredible foresight.

Even if the dark energy is Einstein's cosmological constant, however, there are still problems. Physicists are at a loss to explain exactly where it comes from, or why it has the particular strength that is has. (Their best efforts to predict dark energy's strength, based on what we know about subatomic particles and quantum theory, gives a value that's far too large, by many orders of magnitude.) The nature of dark energy remains one of the most profound mysteries in physics today.

Dark Energy, Dark Future

One thing we do know about dark energy: the extra "push" it delivers would seem to guarantee an open, ever-expanding cosmos. Today astronomers look out across the universe and see galaxies lumped together in clusters, with those clusters grouped together in "super-clusters." The superclusters, in turn, appear to be strung out in vast string-like filaments that stretch for hundreds of millions of light-years across the cosmos. Gravity has crafted these structures – but dark energy will tear them apart.

As fate would have it, Adams published *The Five Ages of the Universe* just before Schmidt, Riess, and Perlmutter announced their discovery. How does the presence of dark energy affect Adams's forecast?

"Perhaps the most important update is that we now 'know' that the universe is accelerating," he told me by e-mail (using the quote marks to emphasize the fact that, in science, no result is ever 100 per cent certain). "Since the expansion of the universe is speeding up, essentially no more cosmic structure will form." In other words, those clusters and superclusters and stringy filaments are the end of the line in terms of cosmic evolution. "The things that we have now in the universe will be all that you get – ever," he says.

Thanks to dark energy, those large-scale structures will gradually disintegrate, and the universe will eventually look very different from what we see today. Things will appear fairly normal for the first few trillion years; stars will continue to shine, and any planets they may harbor could be reasonably hospitable places. Adams calls this the "stelliferous" era (the term means "filled with stars"); it is the era we now inhabit.

Eventually, however, the stars will exhaust their nuclear fuel, and – perhaps 100 trillion years from now – no new stars will be able to form. The stelliferous era will have come to an end, and we will enter what Adams calls the "degenerate era": the most prominent objects in the universe will be "degenerate stellar objects" – essentially the wasted cores of stars that no longer shine. Ordinary stars will have evolved into white dwarfs, while heavier stars will become ultra-dense neutron stars or black holes. (Very occasionally, two white dwarfs will collide, triggering a supernova explosion. In what is left of our galaxy, Adams has calculated, this will happen once every trillion years or so. Each supernova will shine brilliantly for a few weeks but in the end will decay into degenerate cores along with every other star-like object remaining in the universe.)

Ultimately, we shouldn't get too attached to these stellar remnants either, Adams cautions. After a mind-numbing period of time, white dwarfs and neutron stars will disintegrate through a process called proton decay, in which solid matter gives way to radiation. (The lifetime of a proton is not yet known, but our best theories suggest that protons last for 10^{30} to 10^{40} years.)* This marks the end of the degenerate era. After that point, the only sizable objects left in the universe will be black holes, and we enter the aptly named "black hole era."

Black holes are the most enduring objects that our universe and the laws of physics are able to craft. And yet they, too, must succumb to the endless time of an expanding universe. Black holes will ultimately disappear, evaporating by a process known as Hawking radiation (a quantum-mechanical process first described by Stephen Hawking in 1974). A black hole with the mass of the sun may last for 10^{65} years; a supermassive black hole may endure for 10^{100} years. (That number may look familiar: it's called a googol – a one followed by one hundred zeros.)

---

* These are very large numbers indeed. Remember that the age of the universe at the moment is only about 10^{10} years.

After the last black hole has disappeared in a puff of Hawking radiation, the universe will be nearly empty. All that will remain is a sparse flotilla of fundamental particles, drifting endlessly across a frozen, featureless void. Adams calls this final epoch the "dark era."*

If we could somehow transport ourselves to this distant era, what would we see? "Very little," says Adams. "The universe would be very, very dark, very diffuse." All that will remain is a "very diffuse 'soup' of particles. Mostly elementary particles: electrons, positrons, neutrinos, and photons – and perhaps other things that we don't know about." Not much can happen in this rarefied environment, Adams explains. Occasionally, an electron will bind with a positron to form an atom of "positronium" – but even these will eventually disintegrate. Electrons and positrons can also directly annihilate with each other. "And except for these low-level annihilation events," says Adams, "the universe is a very low-energy, low-key kind of place . . . A sea of darkness."

It is perhaps T.S. Eliot rather than Robert Frost who came closest to the mark: "This is the way the world ends / Not with a bang but a whimper."

The End of Astronomy

It is hard to think of anything more depressing than this slow decline of the cosmos into eternal darkness. But here goes: because of dark energy's unforgiving push, the night sky of the remote future will be far less rich than the one we see today, and astronomers of that era will have no inkling of the vast and complex cosmos that once existed.

The Milky Way and its closest neighbor, the Andromeda Galaxy, are bound together by gravity; together with a sprinkling of "dwarf" galaxies,

---

* I have mentioned four of the "five ages" that provide Adams with the title for his book. Our current age, the stelliferous era, is the second; it was preceded by the "primordial era," which covered roughly the first million years of cosmic history, from the big bang to the creation of the first stars.

they make up the so-called Local Group. The billions of other galaxies beyond the Local Group are not gravitationally bound to us, and the expansion of the universe, driven by dark energy, will eventually push them out of view. The most distant objects will be the first to disappear – "cloaked behind a cosmological horizon," as Adams puts it. Nearer galaxies will follow, slipping away one by one.

By 100 billion years from now, give or take, even the Virgo cluster – the next-closest cluster of galaxies beyond the Local Group – will have disappeared over this cosmic horizon. We will be completely isolated from the rest of the universe: beyond the handful of galaxies that make up our Local Group, our telescopes will reveal only blackness. All of those other clusters will suffer the same fate; each of them will be similarly isolated from *its* neighbors. Should astronomers exist in those other realms, their telescopes, too, will reveal nothing. Kant's vision of "island universes" will have been realized in the most literal way.*

Our Local Group will still see some action: our galaxy and the Andromeda Galaxy are currently moving toward one another, and they are expected to merge in about 6 billion years. (The merger will not directly affect most stars: because stars are very far apart compared to their individual diameters, a typical star will not itself undergo a collision.) In the long run, the Milky Way, Andromeda, and the other smaller galaxies of the Local Group will merge into one large conglomeration.

When our Local Group becomes a universe unto itself, astronomers will have things to aim their telescopes at "locally" but will be ignorant of the universe's overall structure. As Lawrence Krauss and his colleagues have recently argued, astronomers of this distant era will be hard pressed to infer that anything like a big bang had ever occurred: with those distant

---

* One way to think of this disappearing act is that these galaxies will be receding from each other faster than light can travel the distances that separate them. (This sounds like a violation of special relativity, but it isn't: it is the expansion of space itself that is driving the galaxies apart.) Equivalently, we can think of their light as being so severely red-shifted as to be undetectable.

galaxies red-shifted into oblivion, it will no longer be possible to make the discovery that Hubble made in the 1920s. The cosmic microwave background radiation, meanwhile, will suffer a similar fate: as the CMB radiation gets stretched out to longer wavelengths, its signal will be lost among radiation from other sources. Krauss says that astronomers living at that time will be misled: "It will lead them to the wrong conclusion about what the universe is doing. The universe will look static, and that's vastly wrong, because the universe is expanding so fast they can't see it."

This is troubling on many levels. It is naturally disheartening to think of knowledge that we have today no longer being available in the remote future; perhaps it will make us strive to preserve that knowledge at all costs. It may also make us wonder just how confident we should be in our interpretations of what *we* see in the sky right now. On the other hand, there has got to be a compelling science-fiction tale in all of this: *Civilization A proudly declares that they have mapped the entire universe, only to be confronted by Civilization B, whose dusty records of long ago describe a night sky that told of a more wonderful and infinitely larger cosmos, now lost . . .*

The End of Life

We have seen how the universe is destined to end in darkness; what, then, is the fate of life in the universe? The second law of thermodynamics, once again, seems to dictate our destiny. In an open universe, it would seem that every entity, every being, every thought, must come to an end. As the philosopher Bertrand Russell once put it, "All the labours of the ages, all the devotion, all the inspiration, all the noonday brightness of human genius, are destined to extinction . . . The whole temple of Man's achievement must inevitably be buried beneath the debris of a universe in ruins."

In the late 1970s, however, Freeman Dyson suggested a way out: he imagined "life" in more general terms as anything that can process information. Because information processing requires energy and generates heat, it would seem as though an expanding universe offers less and less usable energy to keep such a system functioning. Dyson suggested that life could in effect "hibernate" for ever-increasing periods of time. By lengthening the span of the hibernation periods – effectively lowering their "metabolism," so to speak – life could endure more or less forever, he asserted.

With the discovery of dark energy, however, Dyson's strategy may no longer work. Physicists Lawrence Krauss and Glenn Starkman considered the problem in the late 1990s and found that life is indeed in trouble. Life requires energy, they reasoned, and in an accelerating universe, it becomes more and more difficult to collect and harness that energy. As we become isolated in our respective "island universes," the resources at our disposal become strictly limited. With finite resources, any living creatures (or equivalent machines) would have a finite memory, they argue, and "would eventually have to forget an old thought in order to have a new one." Finite information, they argue, implies a finite number of thoughts. In the end, thinking entities would be left with little to do but to have the same thoughts over and over again. "Eternity would become a prison, rather than an endlessly receding horizon of creativity and exploration." In the long run, "life, certainly in its physical incarnation, must come to an end."

This is not a particularly happy outlook for life, the universe, and everything, but perhaps we can take away something positive from our speculation. First of all – and I seem to recall Carl Sagan saying something like this toward the end of his TV series *Cosmos* – all those billions of years that lie ahead offer the opportunity to do a great deal of good. Further, it is quite impressive that with our finite hominid brains we have been able to peer so far ahead, with at least some degree of confidence. It is also rather intriguing that the fate of the universe billions upon billions of years in the future is actually clearer to us than the fate of our own civilization just a few centuries from now.

Perhaps our descendants will congregate in the Nevada desert and stand in awe before the Clock of the Long Now, in the same way that tourists today stand humbled before the pyramids of Egypt. Or not. It is anybody's guess if the fruits of this ambitious project will endure for so many generations. Physicist and writer Gregory Benford, for one, suspects the grand ten-thousand-year clock may meet its demise on a much shorter time scale. The prototype clock is "too pretty," he says. He hopes that the full-sized version planned for the desert is less shiny. "The first biker gang to come along is going to trash this thing," he says. "If they had made it rugged and ugly, it might last."* Writer Brian Hayes is even more critical: by assuming that civilization ten thousand years from now would share any of our values, let alone our desire to keep track of time, we're committing an act "of chronocolonialism, enslaving future generations to maintain our legacy systems." Hayes acknowledges that it is noble to act in the best interests of posterity – but wonders how we might guess what those interests would be, beyond a few generations. "To assume that the values of our own age embody eternal verities and virtues is foolish and arrogant," he writes. "For all I know, some future generation will thank us for burning up all that noxious petroleum and curse us for exterminating the smallpox virus." Even the five-digit coding of the years, which seems like a good way to avoid a "Y10K" crisis on New Year's Eve, A.D. 9999, is wrong-headed, Hayes argues. Four digits, he says, is plenty. "If we take up the habit of building machines meant to last past 10000, or if we write our computer programs with room for five-digit years, we are not doing the future a favour. We're merely nourishing our own delusions."

Indeed, Hayes notes that in centuries past, other clocks were built with lofty hopes that they would run for equally lengthy periods. He cites the example of the grand astronomical clock in Strasbourg Cathedral,

---

* In his book *Deep Time* (1999), Benford gives another startling example of our failure to communicate across time: many "time capsules" that have been sealed (and often buried) by well-meaning citizens, for the interest of future generations, have since been lost, either by "markers" not being erected or by the locations of the capsules simply being forgotten. He notes that the city of Corona, California, has laid down seventeen time capsules in the last fifty years – and has lost track of all of them.

originally crafted in the 1300s. Two centuries later, when a team was brought in to repair the clockwork, they instead chose to start from scratch. The clock was overhauled again in the 1700s, and once again workers installed a new mechanism instead of repairing the previous machine. Hayes suspects that the Clock of the Long Now, ambitious as it is, will be similarly overhauled and replaced long before its ten thousand years are up. At the very least, the full-scale version will have to be different from the London prototype in one very important respect. I said earlier that the prototype "ticks twice a day," and that its three-pronged pendulum "rotates back and forth" – but I should really have said that the clock *would* tick, and the pendulum *would* swing, if the clock were actually running.

Apparently the clock was switched off when it was shipped from California to England, and since being installed in the Museum of Science it has yet to be switched back on. It was inert at the time of my visit, and remained so as of early 2008. Part of the problem, a spokesman for the Long Now Foundation says, is the glass case that now surrounds the clock, making it difficult to wind the mechanism. The goal is to come up with a motorized winding/driving mechanism that can be maintained without opening the case, he says. It's not quite clear how long this upgrade will take.

Until then, the clock designed to speak across the ages remains silent.

ILLUSION AND REALITY

Physics, philosophy, and the landscape of time

Time is a river which carries me along, but I am the river; it is a tiger that devours me, but I am the tiger; it is a fire that consumes me, but I am the fire.

– Jorge Luis Borges

Time is an illusion. Lunchtime doubly so.

– Douglas Adams

On our tour so far, we have glimpsed time from a number of different angles. We have met those who see time as absolute and those who see it as relative; those who see it as a line and those who see it as a block – and even those who would like to fold it back on itself and travel through it in a loop. We imagined the lives of those who first became conscious of time's passage and those who have learned how to slice each second into a billion parts. We have looked at the beginning and end of time – or at least peered as closely as science is currently able to allow us.

And yet some of the most basic questions about the nature of time remain unanswered. To begin with, there's that pesky issue of time "flowing." Does time truly "pass by" in some tangible way? It is an ancient question, one that begins in earnest with the conflicting views of Parmenides and Heraclitus; one that has troubled the greatest minds from Augustine to Newton, from Kant to Einstein. Is time nothing more than change? Or is time more fundamental – is it the mysterious entity that *makes change possible*, a kind of foundation on which the universe is built? Or is it just the opposite: as much as we like to speak of the "river of time," could the river be dry, its flow an illusion? (And how *can* it flow if it is meaningless to speak of the *rate* at which it flows?) If the flow is imaginary, have "past" and "future" dried up along with it, leaving only an array of "nows," all on an equal footing, as Julian Barbour and other daring thinkers have suggested?

Perhaps millions of years of biological evolution, coupled with thousands of years of cultural and linguistic evolution, have shaped our minds in such a way that we imagine such a flow where none exists. The problem then becomes one for psychologists and philosophers. Perhaps it will turn out to be as complex and difficult as the other great problems in those fields – problems like "What is the self?" and "What is consciousness?"

Between Mind and Brain

Is the passage of time something our brains assemble out of a swirl of sensory data and then present to us as though it were real? Is the process so efficient, perhaps, that we imagine that the finished product was "out there" all along? For some thinkers, the "self" itself is such a construction – in which case time might simply be one facet of a much richer cognitive assembly.

"In essence, the self is a construction of the brain," observes philosopher Patricia Churchland of the University of California in San Diego. The self is "a real, but brain-dependent organizational network for monitoring body states, setting priorities and, within the brain itself, creating a separation between inner world and outer world." A simple example of this, she says, is visual perception. We "see" two-dimensional images of the world – one from each eye – but our brains take this data and build a three-dimensional image, and that is what we perceive. "The brain constructs a range of make-sense-of-the-world neurotools," Churchland writes. "One is the future, one is the past and one is the self." She emphasizes that this does not make the self – or the past or the future – unreal; but she sees them as "tools" that we use, rather than aspects of the world.

Physicist and author Paul Davies seems to agree. I met with him recently in Phoenix, where he is the director of a new research center called Beyond, on the campus of Arizona State University. (As hot as New York is in July, Phoenix in September is worse. It's the only campus I've visited where they've installed overhead sprinklers in front of the bookstore to provide brief moments of cooling between classes.) In 1995, Davies wrote a very thorough and insightful book about the physics of time, called *About Time*. In that book, and in numerous essays and articles, he argues that the flow of time is an illusion – and he held fast to that stance during our interview.

"There is nothing that corresponds to a flow of time, a movement of time, a past, present, and future, or a now," he told me. "None of that is there in physics." As a result, he says, it is "very tempting to believe that this is just a product of psychology and language [and] has nothing whatever to do with the nature of the physical world."

We take the flow of time as "common sense" – but our common sense, along with our intuition, is a product of biological evolution, Davies says. "Evolution has built our minds like it's built our bodies. So we're good at thinking about the world in certain ways, and we're comfortable with certain concepts." And it seems we have become *very* comfortable with the idea of time as a thing that passes by. As a result, he says, concepts like the flow of time – even though they have no basis in physics – have come to "form part of our picture of the world."

I wonder if that's what Augustine was thinking in the fifth century A.D., when he described the flow of time as something perceived not in the world but in the mind: "It is in you, my mind, that I measure time . . . As things pass by, they leave an impression on you . . . It is this impression which I measure. Therefore this itself is time or else I do not measure time at all." Immanuel Kant, writing thirteen and a half centuries later, seems to have agreed: "The idea of time does not originate in the senses but is presupposed by them," he declared. "Time is not something objective. It is neither substance nor accident nor relation, but a subjective condition, necessary owing to the nature of the human mind."

If the flow of time is an illusion, it would hardly be the only trick that our brains have played on our minds. Davies uses the example of a child who spins around on a rotating office chair: the child whirls around and around, and then stops – and then feels as though *the room* is spinning. But the child knows it is not so: it is an illusion, and in a few minutes it passes. The flow of time, Davies says, may be the same kind of illusion – only far more deeply ingrained.

But we still want to know *how* that appearance of "flow" comes about. "This illusion cries out for an explanation," Davies has written. And that explanation, he suggests, "is to be sought in psychology, neurophysiology, and maybe linguistics or culture."

The Evolution of Time

There have, of course, been educated guesses. We have seen how Roger Penrose has grappled with the links between time's various arrows. Robert Jaffe, a physicist at MIT, suspects there is a connection between the thermodynamic arrow of time and the psychological arrow. "My impression is that the human experience of time is fundamentally thermodynamic in origin," he told me recently. "Our perception of memory and experience and anticipation of the future – all of those things have their origin in information storage, processing, deterioration, and entropy production in the physical-chemical environment which is the brain."

The brain does indeed move information around; whether it does so in a fashion similar to that of a computer has been endlessly debated. (Penrose, for one, has argued passionately *against* that view.) Recently there has been a flurry of arguments suggesting that information theory will one day help us understand the laws of physics as well as the nature of consciousness.* This is highly speculative, but if it can shed any light on the origin of time's apparent passage, it is worth pursuing.

Others approach the problem from the perspective of evolution. Our remote ancestors certainly did not encounter time's flow in quite the same manner that we do; the way we perceive the world has evolved over time. It developed as our brains developed. We are bombarded with raw information from our environment at every moment – shapes and colors, shades of light and darkness, sounds and smells and so on – and our brains somehow integrate that information to forge a coherent picture of the world. Without that ability, we would be numbed by the chaos of that data stream itself. Instead, we construct a "scene." The survival value is obvious: we don't see a pair of gleaming eyes and a bunch of black and yellow stripes; we see "a tiger" – and we run.

Is the flow of time a similar construction? James Hartle, the physicist who worked with Hawking on the "no boundary" proposal (Chapter 10), is among the growing number of thinkers who has come to that conclusion. "Our powerful sense that there is a 'now' and that time 'flows' from the past, through the present, to the future, has survival value," he writes. He is also hopeful that if we establish the constraints that evolution has placed on our perceptions, it could help us understand the true nature of time – with a little help from information theory. His idea, to summarize very roughly, is that any entity that gathers and uses information (this would include human brains) initially holds that information in "input registers," but then transfers it to "memory registers" to free up space in the input areas. That passage of information between registers somehow "feels" like time. "Something analogous to the flow of information from register to register happens in our brains," Hartle says, "and this is what ultimately gives us our sense of time passing."**

---

* Two stimulating books exploring this idea were published in 2006: Charles Seife's *Decoding the Universe* and Seth Lloyd's *Programming the Universe*.
** One implication, Hartle says, is that any extraterrestrials that we encounter will probably have a conception of time similar to ours, "sharing concepts of past, present and future, and the idea of a flow of time."

Jaffe, meanwhile, says that although we do not yet know the details of the actual mechanism, it is "probably an incredibly complex, integrative process." Our feeling that time flows is "a deeply innate evolutionary adaptation to the chaotic experience that these DNA-carrying organisms" – our remote ancestors – "found themselves in. We've inherited it from the very earliest stages of natural selection."

Seeing and Believing

Nature presents us with many kinds of illusions. For millennia we gazed at the stars of Orion, and we saw a mighty hunter. Today we can relish in the mythology represented by that vision, and we can still enjoy the majesty of those brilliant stars on a winter evening – but we now recognize that the hunter lives only in our imaginations. We have also shattered many related illusions: that the world is flat, for example; that the sky is a vast dome; that the sun moves around the earth. These and many similar illusions stem merely from the fact that we observe the universe from a particular vantage point – "illusions of perspective," we might call them.

There have also been what we might call "illusions of interpretation." For centuries we marveled at the plants and animals that surround us, and we saw what seemed like carefully designed structures; their workings made us think of finely crafted machines – perhaps clocks – and so we imagined a divine clockmaker was responsible. But then Darwin showed us how complexity could arise through natural processes, and we realized the invisible clockmaker was unnecessary.

A third kind could be called "illusions of emergence." Consider the "wetness" of water, mentioned earlier. We now know that this wetness emerges only when we have millions of water molecules together; individual water molecules have no such property. (One might also argue that a sentient observer is needed to *experience* that wetness.) The "hardness" of matter is a similar illusion: atomic physics has shown us that a rock is mostly empty space, but we are oblivious to the vast tracts of nothingness between the nuclei of atoms. To our macroscopic touch, a rock is – well, rock-hard.

To greater or lesser degrees, we have learned to live with these illusions. It has taken centuries (perhaps millennia), but today we recognize them for what they are, and we move on. We still teach our children the constellations (or we would, if we knew them ourselves, and were lucky enough to have access to a dark sky), but a child today understands that the hunter in the sky is not real. We may speak metaphorically of the "design" of the human eye, but we know it is not like the design of an automobile. And water without wetness is hardly a problem: remove the wetness, and we still have the jostling water molecules. This is not a devastating blow to our understanding of water – only a gentle reminder that our perception of "wetness" is contingent on many things (and that it is, above all, a perception).

If the flow of time is an illusion, it seems to be more profound than any of these. Physicists may tell us that time "emerged" at the moment of the big bang – but it seems to be a more confounding emergence than that of water's wetness or a rock's hardness. Taking the "flow" out of time seems more disturbing than taking the wetness out of water. Without time's flow, what are we left with? Is time still recognizable as . . . *time*?

The idea that time is not a fundamental part of the universe is a challenging notion. But then, the last hundred years of scientific inquiry have thrown many such challenges our way, beginning in the early decades of the twentieth century, when we had to abandon Newton's absolute time and space in favor of relativity and quantum theory. Four-dimensional spacetime is hard to imagine. (Four-dimensional *anything* is hard to imagine.) Cats that are alive and dead at the same time, and particles whose properties are mysteriously correlated with other distant particles, are hard to imagine. And time not being a fundamental aspect of nature is *very* hard to imagine. As Brian Greene laments, "Whenever I sit, close my eyes, and try to think about things while somehow not depicting them as occupying space or experiencing the passage of time, I fall short. Way short. Space, through context, or time, through change, always manages to seep in." One is reminded of the words of Aristotle, who expressed a similar concern twenty-three centuries earlier: "Even when it is dark, and we are not being affected through the body, if any process goes on in the mind, immediately we think that some time has elapsed too."

Philosophical Interlude, Part Three

One of the most intriguing features of this peculiar dimension is that it has never been quite clear which discipline is "in charge" of studying it. We have already seen how the struggle to make sense of time has involved not only physics but also psychology, linguistics, anthropology, neuroscience and cognitive science, and of course philosophy. And even those disciplines are not cleanly divided from one another; as we have seen, there are specific problems – like the evolution of memory, for example – that naturally embrace more than one field of study.

In this book I have focused largely on physics – but there is also a branch of philosophy known as *metaphysics* which occasionally brushes up against (and perhaps overlaps with) traditional physics. Questions about the ultimate nature of reality, the nature of the mind, and the origin of the universe are often described as metaphysical problems – although, as we've seen, scientists have also investigated these matters, and have made remarkable progress, especially (as we saw in Chapter 10) with the problem of how the universe began.

Time, it seems, is fair game for investigators from all fields. Perhaps physicists have made the most concrete progress, but no one is excluded from the discussion. Particularly in philosophical circles, the debate rages on as vigorously as ever. The "arrow of time," the perceived "flow" of time, and the very nature of time all seem to be up for grabs.

In the philosophy journals one can read article after article on problems that go back to Aristotle (and sometimes further) and yet are presented with some new technical twist or rhetorical spin that manages to make the most ancient problems seem new and urgent. Compared to the often mechanical style of writing seen in physics journals, the pages of the philosophy journals often seem downright passionate.

In the last few years, the prestigious British journal *Philosophy* has run many intricate, argumentative, and even hostile papers by some of today's leading thinkers on the nature of time. In one heated exchange, Michael Dummett of Oxford battles Rupert Read of the University of East Anglia. Dummett gets the ball rolling with a paper titled, "Is Time a Continuum of Instants?" in which he tackles a problem that has troubled thinkers from Aristotle and Augustine onward. Read responds by asserting that "it is unclear how one can possibly build up time, as a continuum, from durationless instants. It is very like the notion of building up a line from dimensionless points." (His paper has a wonderful title: "Is 'What Is Time?' a Good Question to Ask?") Dummett then lashes out at Read, saying that he "appears not to understand" what a continuum is, and cites the set of real numbers as a counterexample.* A few pages later he says, "Dr. Read appears to go berserk, lashing out with wild haymakers, unaware of where his opponent is standing." In yet another rebuttal, Read says Dummett "does not take seriously *enough* the ineluctable sense in which time is *conceptual*, and not simply something which we *find* in (the fabric of) the universe." Who knew that philosophy was so much like fencing?

Whether physicists have any interest in such squabbles is another matter. As far as I can gather from speaking with numerous physicists over the years, most of them are not particularly interested in the philosophical side of the debate. (Keeping up on the literature of their own highly specialized fields no doubt consumes enough time by itself.) As Read, the philosopher, concedes, "In physics, 'time' is whatever it is, and philosophy will never second-guess physics successfully or even, I suspect, remotely usefully."

---

* The *real numbers* are all the numbers on the "number line" – not just the integers, but also the infinity of fractional numbers in between each of them. (That includes "irrational" numbers such as π.)

Yet philosophers, perhaps better than physicists, have a knack for stepping back and struggling to take in the big picture. In his *Treatise on Time and Space* (1973), philosopher J.R. Lucas wrestled with all of the complexities that make up the problem of time. He understood very clearly the inherent contradictions that come to the surface when we ponder something as elusive and enigmatic as time: like Augustine, we think we know what time is, but we work ourselves into a mental and linguistic frenzy as we try to express it in words. We "listen with respect," Lucas writes, when we are told that time is a moving image of eternity, a measure of change, an extension of the mind, the order of events, the readings given by clocks, or the fourth dimension. "But although we listen with respect, we do not give our wholehearted assent." None of these definitions, he says, captures the essence of time. "We cannot say what time is," he laments, "because we know already, and our saying could never match up to all that we already know."

Does Caesar Live?

For most of us, relegating the flow of time to mere illusion would require a novel and counterintuitive way of thinking. We have already seen some of the disturbing problems it presents – such as the alleged negation of free will. But perhaps there is also a positive side to a universe in which every "now" has the same status. According to some thinkers – including Julian Barbour, as we saw in Chapter 6 – it offers a certain kind of immortality.

Why do we feel sad when a loved one dies? There are many answers we could give to such a question, but ultimately our distress comes from thinking that the person no longer exists; that their triumphs and tragedies now lie in the past, while the future simply excludes them altogether. As the philosopher Michael Lockwood has put it, we feel this way "because we instinctively equate existence – full-blooded existence – with existence *now*, at the present moment."

In the tenseless, now-less block universe of modern physics, death can be seen in a new, significantly less tragic light. Past events are said to be just as real as present events, so a life that has "ended" has, in some sense, merely been relocated to a more remote section of the "block." (Indeed, in this view, a life "ends" only in the sense that Arizona "ends" at the New Mexico border; but Arizona has not "gone" anywhere at all.) "From this perspective," writes Lockwood, "a person who is not living *now*, but did or will live at other times, exists in just as substantial a sense as someone who does not live *here*, but only at some other place . . . Regarded in this light, death is not the deletion of a person's existence. It is an event, merely, that marks the outer limit of that person's extension in one (timelike) spatio-temporal direction, just as the person's skin marks the outer limit in other (spacelike) directions." If we take this perspective seriously, then a loved one who has died must be considered "still" alive, just as a friend who lives in a distant country is alive but (in the absence of phones and e-mail and such) out of reach. "Einstein is urging us to regard those living in times past, like those living in foreign parts, as equally *out there* in space-time, enjoying the same flesh-and-blood existence as ourselves," Lockwood writes. "It is simply that they and we inhabit different regions of the continuum."

We have already seen how Barbour embraced this sort of immortality – if that is the right word for it – with his conception of "Platonia," a new way of imaging time in which every "now" simply *is*. During our conversation in South Newington, I asked him if Julius Caesar was just as alive as either one of us. Yes, he said. I tell him I disagree: Caesar is dead; his body decomposed long ago, and that's the end of it. A religious person might say that his "spirit" lives on, or words to that effect, but he is not "alive" in the usual sense of the word.

Barbour stands firm.

Worried that we may be floundering on semantics, I ask him if he believes that Julius Caesar is "literally alive": Whatever it is that *I* possess that makes *me* alive, does Caesar have those qualities too? "I would say yes," Barbour replies. "And I would say that you and I, at what we would regard as earlier stages in our life, are to be regarded as just as much alive as we are now."

I would like to believe him.

I try to imagine the twelve-year-old me still "existing" – enthusiastically trying out his (my) first camera, a Kodak Instamatic given by a generous grandfather . . . or the twenty-one-year-old me struggling through fourth-year quantum physics, with its cryptic equations . . . but I can't quite do it: "they" are gone, and "I" am here instead. And I have infinitely greater difficulty imagining the fifty-year-old me or sixty-year-old me as "existing"; their time will (hopefully) come, but only "I" seem to exist "now."

Mind you, if we accept Barbour's vision – that all of the past and future "yous" are just as real as the "you" who is reading this sentence – then *everyone* who ever lived is still "alive." From this perspective, Hitler and Stalin are just as "alive" as anyone else. It seems to be a double-edged sword: if we are comforted to think of a loved one as still being alive (somewhere in spacetime, or Platonia, or whatever we call this now-less arena), we must also be distressed to think of every instance of pain and suffering and injustice as also still being underway. Barbour doesn't like that phrase – to say something is "underway" or "happening" or "going on" suggests, misleadingly, that time is passing. He prefers to say that those events just "are."

This is a remarkable view of existence. It is hard to grasp. I think I understand it, but I don't think I can subscribe to it. Can Barbour? Surely he must have this feeling – that time flows, that it passes – that the rest of us do?

Yes: he admits he is no different from anyone else on that score. "I live more or less as other people do," he says. Even so, he may have benefited from entertaining these unorthodox ideas about time. "I think my theory has made me savor the instant, perhaps, more than many other people," he says. He tells me he doesn't worry about the "rat race" – indeed, South Newington is about as far removed from the urban rush as one can imagine. If there is anywhere one can live without encountering the stresses of the modern world (at least in England), this is the place. "And I do think the world is extraordinarily beautiful and interesting," he adds. "And I want to appreciate it as much as I can."

Talking Time in Toronto

Barbour's ideas are quite far from the mainstream – but he does have his share of followers. At the very least, he has inspired a new generation of physicists who respect and admire his ability to think about old problems in new ways.

One of those admiring physicists is Lee Smolin of the Perimeter Institute for Theoretical Physics in Waterloo, Ontario (we met him briefly in Chapter 6). The Perimeter is an independent research institute about an hour's drive from Toronto, where Smolin lives. He is perhaps best known as one of the creators of "loop quantum gravity," a proposed alternative to string theory which aims to unify relativity and quantum theory. According to loop quantum gravity, space and time may be quantized (as in Penrose's twistor theory); the term "loop," as Smolin puts it, "arises from how some computations in the theory involve small loops marked out in spacetime." One can even think of space (and therefore time) as in a sense composed from these loops, which would be so small as to be almost unimaginable – perhaps 10^{-35} meters across.[*]

More recently, Smolin is known for writing a rather provocative book, *The Trouble with Physics* (2006), seen by many as an attack on string theory. In the book, Smolin argues that string theory has promised the world but, after decades of hype, has failed to deliver. He also says that too many bright young physicists are joining the string bandwagon, at the expense of other possible approaches. It did not make string theorists happy. (Smolin emphasizes that his book is not an attack on any particular string researchers.)

Smolin has a Ph.D. from Harvard, and taught at Yale and Penn State before settling in Ontario. He lives on a quiet residential street in Toronto, about halfway between the old "happening" strip of Queen Street West

---

[*] In *The Fabric of the Cosmos*, Brian Greene sums up the competing theories as follows: "A one-sentence comparison would hold that string theorists start with the small (quantum theory) and move to embrace the large (gravity), while adherents of loop quantum gravity start with the large (gravity) and move to embrace the small (quantum theory)." Of the two approaches, he says that string theory has made greater progress.

and the newly popular strip known, for lack of a better name, as West Queen West.

Smolin tells me he greatly respects both Penrose and Barbour. He has said in the past that Barbour, in particular, has been a "philosophical guru" for him. He especially admires Barbour's approach to the problem of quantum gravity; many who have tackled the issue have displayed "sloppy thinking," Smolin says, while Barbour has "really thought it through."

But Smolin does not follow Barbour all the way to the Englishman's "timeless" conclusion. Smolin is not willing to recognize all "nows" as equal. Such a move simply takes too much out of time's essence; what remains would not be "time" as we know it. "On the philosophical side, I believe that time really is fundamental; I can't get that out of my head," Smolin tells me. "Time is so fundamental in our experience of nature; how could it not be a fundamental part of the world?"

Smolin, fifty-two, has somewhat unruly hair and a graying beard; he wears wire-rimmed glasses and sports a black fleece sweater with the sleeves rolled up. As we sit at his large-dining room table, a variety of people (and one animal) pass through: the housekeeper; Smolin's fifteen-month-old son, Kai; a large black dog named Emily; and the man who sometimes walks her. (His wife, a lawyer, is not home at the moment.) "I just hang out here," Smolin remarks. "People come and go."

With all this coming and going, how could anyone deny that time passes?

Physicist Lee Smolin. *(© Dina Graser)*

"We experience reality as a succession of moments," says Smolin as we sip tea and munch on chocolate chip cookies. "It's difficult to characterize, but in any case it's clear that there is a 'present' – however confusing it is to talk about it – and there was a 'past' and there will be a 'future.'"

That sounds reasonable enough; not long ago I would have accepted such statements as self-evident truths, even rather obvious ones. Indeed, they *still* sound true.

"As we sit here and talk, the present is always new; what *was* the present no longer exists," Smolin continues. "And this is absolutely fundamental. [It is] the most irreducible feature of our experience of reality."

In other words, like Penrose and Barbour, Smolin believes that physics has not paid enough attention to time. Our theories – as good as they are – have failed to capture something rather essential about its nature.

For Smolin, time – including that troublesome "flow" – is simply too real to dismiss as an illusion. And therefore Barbour's vision of a timeless world is one he cannot embrace.

I mention my exchange with Barbour about whether Julius Caesar is alive or dead. "I disagree with Julian," Smolin says. "Caesar no longer exists."

Nor is he satisfied with the snarky-sounding definition sometimes given by physicists – that time is simply what clocks measure, nothing more. That may serve well enough as an operational definition, but it is also "a kind of operational cop-out," he says. "And I'm not an operationalist about these things." For one thing, such a definition seems to ignore cause and effect, along with that oft-mentioned flow.*

---

* Another "escape route" is offered by the philosophical stance known as *positivism*, which focuses on the results of measurement and observation rather than attempting to discern "reality." Stephen Hawking, for example, has stated, "If one takes the positivist position, as I do, one cannot say what time actually is. All one can do is describe what has been found to be a very good mathematical model for time and say what predictions it makes."

Smolin had expressed a similar concern in his book *The Life of the Cosmos* (1997): "Speaking personally, my imagination quails before a world without change and time. I don't know if there are any real limits to what the human mind can imagine, but thinking about this question brings me closer than I like to the limits of what my own mind has the language or the means to conceive."

I bring up the suggestion that the flow of time is a property of the mind and not a property of the universe; that the ultimate explanation for time's passage may be psychological.

"I've puzzled over that a lot," he says. "I don't see how there could be such an explanation."

The notion of time "flowing" was already slipping away before Einstein came along. We might say that it began to erode with the work of Galileo, Descartes, and Newton – the towering figures of the Scientific Revolution who took the first steps toward a geometrical picture of time (even though it was Newton who mentioned that "flow" in his famous definition). Since then, we have become very used to the idea that time is a line that we can plot on a graph *just as we represent a direction in space*. If time and space are plotted on separate axes at right angles to one another, then we can draw another line showing a relationship between them: for example, we can plot a graph of distance traveled versus time for, say, a speeding car.* We ask schoolchildren to do this, and most of them get the hang of it without too much trouble. But as helpful as such representations are, the line the student draws surely does not "flow," nor, for that matter, does it help us to understand such a flow.

We have already seen (in Chapter 6) one direction where this way of thinking may lead us: We can imagine the universe as a kind of "block"

---

* As we would typically plot it, the graph goes up and to the right. In other words, distance increases with time. If the car is moving at a constant speed, you get a straight line.

in which there are many "nows," all of them of equal weight. There is no "flow" transporting them from future to present, or from present to past. And there is certainly no overarching "now" to unite all of us.

This certainly seems to be the direction that Einstein was taking us with special and general relativity, which showed just how intimately linked space and time actually are. As we have seen, with relativity the idea of a universal "now" disappears; instead, we each have our own, and mine is as good as yours. Relativity destroyed any hope for a "universal clock" that we may have had, and threatens to wreak havoc with "past" and "future" as well. Are terms like "past" and "future" rendered similarly subjective?

Many physicists would say yes, end of story. David Deutsch – we met him briefly in Chapter 8 – is one of them. Like Penrose and Barbour, Deutsch firmly believes that the "flow" of time is in our heads. "We do not experience time flowing, or time passing," he wrote in *The Fabric of Reality* (1997). "What we experience are differences between our present perceptions and our present memories of past perceptions. We interpret those differences, correctly, as evidence that the universe changes with time. We also interpret them, incorrectly, as evidence that our consciousness, or the present, or something, moves through time."

With the "flow" of time removed, Deutsch insists, the idea of "now" becomes every bit as subjective as the idea of "here." As we spoke in his suburban Oxford home, I suggested that we draw a "timeline" and mark off the years on it: . . . 2006, 2007, 2008 . . . No problem so far. And then I asked him – it being 2007 at the time of our chat – can't we draw an arrow pointing to "2007" and label it "now"? Only with a caveat, he says: labeling one point on the line "now" makes as much sense as taking a map and labeling one spot on it "here." Labels such as "here" and "now," he explains, are *indexical*: they apply only relatively, not absolutely. "If I show you a world map, I can put a sign on it saying 'here,' and that's exactly the same thing as you're doing when you're labeling one part of [the timeline] 'now.' But nobody would believe that the 'here' is actually part of the world – that there is objectively a part of the world called 'here.'" He adds, "These aren't mysteries, they're just quirks of language."

I can't argue with his logic, but it still feels a little strange. (Although looking back from the perspective of 2008, it is clear that 2007 is *not* "now" anymore – but it *was*, then! Presumably, this was precisely his point.) Still, as you read this sentence, doesn't it feel like there is something special about this moment, right *now*? And doesn't it feel *more special* than the location "here"? Certainly your current location in space is not the same as a spot ten meters to the east or ten meters to the west – but that seems trivial. You *could have* chosen to sit ten meters from where you are at the moment, if you had been so inclined. But it feels as though you don't have any choice at all when it comes to your location in time. You're *automatically* at the moment we call "now."

Deutsch has heard all these arguments many times, and is not moved by them. "You can give a full description of what the world is like without ever mentioning 'here' and 'now,'" he says. (If I were Dr. McCoy to Deutsch's Mr. Spock, this is where I would say, "Confound your logic!")

Why, then, do we have this feeling that time flows? "I don't think we do," Deutsch says. "I think this idea of time flowing is not a *feeling* that we have, it's a *manner of speaking* that we're brought up to use. We don't actually think of the world that way – we just say we do."

Naturally, I can't leave without asking him to weigh in on the status of Julius Caesar. I tell him that I think that I'm more alive than Caesar is.

"Well," he says, "you are *now*."

The Landscape of Time

Many philosophers and physicists – especially post-Einstein – seem content with this state of affairs; they seem to accept a rather static version of time. In an influential essay titled "The Myth of Passage" (1951), the American philosopher D.C. Williams speaks of time as if it is indeed very similar to space. To move through one is to move through the other; both, he insists, can be viewed simply as "ordered extensions." Any "flow" one perceives is just that – a perception – and not something "out there."

"Does this road go anywhere?" asks the city tourist. "No, it stays right along here," replies the countryman. Time "flows" only in the sense in which a line flows or a landscape "recedes into the west." That is, it is an ordered extension. And each of us proceeds through time only as a fence proceeds across a farm: that is, parts of our being, and the fence's, occupy successive instants and points, respectively. There is passage, but it is nothing extra.

The philosopher Hilary Putnam was also satisfied: "I do not believe that there are any longer any philosophical problems about Time; there is only the physical problem of determining the exact physical geometry of the four dimensional continuum we inhabit." And here is the German mathematician Hermann Weyl: "The objective world simply *is*: it does not *happen*. Only to the gaze of my consciousness, crawling up the life-line of my body, does a section of the world come to life as a fleeting image in space which is continuously changing in time."

Well, that clears *that* up, right? Time doesn't flow; instead, it is an "ordered extension"; part of a four-dimensional continuum; an artifact of one's "gaze of consciousness." I must restrain any impulse toward sarcasm: after all, this appears to be the mainstream position in both physics and philosophy today. Whatever time is, it is not something that flows. That is a feature that we read into it, so to speak; it is not something "out there."

But part of me resists. Time flows just as a landscape "recedes into the west"? Nothing of the sort! Landscapes *just sit there*. Time does something quite different: it carries me along. Or it rushes past me. Or something. It is a strange landscape indeed that forces all of us to march through it in lockstep, with no stops or U-turns allowed. Time, as I imagine it, could hardly be *less* like a landscape.

Galileo famously said that nature is written in the language of mathematics – but as Lee Smolin points out, mathematical entities like numbers and lines seem to exist outside of time; they seem to be frozen. As he told a physics workshop in New York recently, "We experience the world in time as a succession of moments. But those moments disappear when we represent the world mathematically." In *The Trouble with Physics*, he asserts, "We have to find a way to *unfreeze* time – to represent time without turning it into space. I have no idea how to do this. I can't conceive of a mathematics that doesn't represent a world as if it were frozen in eternity."

He also suggests that the problem has been many centuries in the making, with each new theory – including the great breakthroughs of the twentieth century – wedding us ever more tightly to a conception of time that is, at root, mistaken. "More and more, I have the feeling that quantum theory and general relativity are both deeply wrong about the nature of time," he writes. "It is not enough to combine them. There is a deeper problem, perhaps going back to the origin of physics."

And Smolin is not alone. Lisa Randall, for example, has expressed her doubts: "I wish time were an illusion," she said recently. "But unfortunately it seems all too real." Even Paul Davies seems to have had moments of uncertainty. In spite of his many essays in support of the tenseless view of time – and in spite of his analogy of the spinning chair – he has also wondered if our description of time is missing something. Toward the end of *About Time*, he says:

> As a physicist, I am well aware how much intuition can lead us astray ... Yet as a human being, I find it impossible to relinquish the sensation of a flowing time and a moving present moment. It is something so basic to my experience of the world that I am repelled by the claim that it is only an illusion or misperception. It seems to me that there is an aspect of time of great significance that we have so far overlooked in our description of the physical universe.

Perhaps the coming decades will bring more satisfying answers to our most pressing questions about the nature of time. Physicists have suggested that "time" as we know it emerged from quantum foam, or vibrating strings, or wobbling branes, or *something* at the moment of the big bang. Fine. But please: more detail. Exactly how did it emerge, and what gave it the properties that we now think it has? And for the psychologists and philosophers: If the flow of time is indeed an illusion, a construct of the mind and the brain, please tell us – how does that illusion come about?

A Stroll through Princeton

Einstein, the man who revolutionized our view of time just over a century ago, was on a working holiday in the United States when the Nazis came to power in his homeland. He was by then acknowledged as the greatest living scientist. He was also, by accident of birth and circumstance, the world's most famous Jew. It was too late for the Nazis to prevent his escape; instead they raided his house outside Berlin and seized his property. Later, Nazi supporters burned his books at a public bonfire. A list of the state's most-wanted enemies showed Einstein's photograph with the caption *"noch ungehängt"* ("not yet hanged").

Einstein immediately renounced his German citizenship, and never again set foot on German soil. He also wrote numerous letters to help secure visas for other Jewish scientists in Germany, saving many lives. In 1933, he accepted an offer from the newly created Institute for Advanced Study (IAS) in Princeton, New Jersey. Einstein settled in an old wooden house on Mercer Street and relished in the seclusion that the bucolic town provided. Princeton would be his home for the final twenty-two years of his life.

One of America's most upscale university towns, Princeton has changed little in the half-century since Einstein's death (give or take a Starbucks or two). Compared to Bern, however, it has done relatively little to celebrate its most famous citizen. Einstein's house – as the scientist had wished – never became a memorial; it remains a private residence. (Though it may still exude genius: the economist who now lives there recently won a Nobel Prize.) On Nassau Street, the local historical society runs a small museum with an array of Einstein memorabilia, and in 2005 – the hundredth anniversary of Einstein's miracle year – a citizen's group erected a bronze bust of the scientist in front of the town hall.

Einstein was acutely aware of the strange new picture of time and space that his theory of relativity had presented to the world, and he may not have been entirely comfortable with it. Nor was he alone in his struggle: his views were shaped, in his later years, by conversations with the brilliant Austrian logician Kurt Gödel (1906–78). Like Einstein, Gödel fled from the Nazis; after crossing Siberia by train, he sailed to America and eventually took up a position alongside Einstein at the IAS in

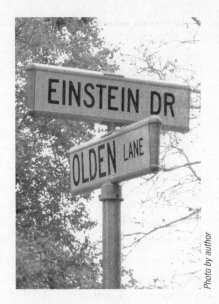

Photo by author

1940. The two men quickly became close friends. One wonders whether they turned heads or were simply ignored as they strolled to and from the Institute every day, discussing the secrets of the universe – in German, of course – while they shuffled along Princeton's leafy residential streets.

The story of their friendship received little scholarly attention until philosopher Palle Yourgrau made it the focus of his recent book, *A World without Time* (2005). In the world of mathematics, Gödel is remembered primarily for his "incompleteness theorems" – a pair of groundbreaking proofs that he developed in the early 1930s that appear to place fundamental limits on the reach of mathematical systems. But as Yourgrau points out, Gödel also pondered matters of physics, including the implications of Einstein's general relativity. He contemplated "rotating universes" described by the theory, and was among the first to worry about "closed timelike curves" and the problems associated with time travel. And he seems to have been particularly troubled by what relativity does to the notion of time. Both Gödel and Einstein wondered what it means for "now" to be subjective – for "now" to be just another "here." But Gödel went further than Einstein: like Parmenides and philosopher John McTaggart, he concluded that time itself cannot be real.

Einstein never disposed of time outright – but he still struggled with the notion of a time that doesn't flow; the removal of a "universal clock" that labeled each "now" in a distinct, unambiguous fashion; a universe in which every event has a past and future, but in which there is no "absolute" past and future. As he describes this mid-century pairing of luminous minds, Yourgrau's own angst is palpable. Can the past and the future be just as real as the present?

> Should I still be wondering what to order for breakfast yesterday, as I am for tomorrow, or should I cancel both orders because the meals have already arrived? And since the present is no more real than the past and I am still lying on the beach as I was last summer, why am I identifying only with the "I" that is presently shivering in the cold? ... Are there as many "I's" as there are moments in time, and if so, are they all me, or only parts of me?

Here is the crux of the problem: relativity makes all moments equal; but to be human is to declare them *un*equal. When Descartes said "I think, therefore I am," he could just as well have said: "I think, therefore I am *now*." Everything about the human experience demands the special attention we pay to the present moment.

Remove now's special status, and the river of time becomes a block, with every part just like every other part. One is led, perhaps, to Julian Barbour's endless array of nows, complete with an endless array of Julian Barbours. And John McTaggart's "B series," in which there are lots of events, but nothing ever "happens." And to the recognition that "now" is just a subjective label, as David Deutsch has asserted. Is this where Einstein's revolution has left us? And, if so, why do we feel as though we inhabit just a *single* now? Will the answer come only when we make sense of human consciousness, as Roger Penrose has suggested? No wonder Yourgrau yearns to know what words passed between Einstein and Gödel on those crisp Princeton evenings half a century ago. "The confidence of the popular (and not so popular) mind is misplaced when it clings to the belief that all is well, temporally speaking, between the universe and Dr. Einstein," he writes. "All is not well at all."

It would be interesting to hear, in his own words, exactly what Einstein thought of such matters. Unfortunately, although he wrote extensively on subjects far and wide – not only physics, but also politics, human rights, religion, and more – we have only hints regarding his thoughts on the disappearance of time. As we have seen (Chapter 7), Einstein referred to "the primitive subjective feeling" of time's flow, a feeling that "enables us to order our impressions, to judge that one event takes place earlier, another later." Another hint – perhaps an important one – comes from the German-born philosopher Rudolf Carnap (1891–1970). Carnap managed to leave Germany soon after the Nazis came to power, and joined Einstein briefly at the IAS before moving to California, where he joined the faculty at UCLA. In an autobiographical essay, Carnap relates a conversation that he had with Einstein in the early 1950s:

> Once Einstein said that the problem of the Now worried him seriously. He explained that the experience of the Now means something special for man, something essentially different from the past and the future, but that this important difference does not and cannot occur within physics. That this experience cannot be grasped by science seemed to him a matter of painful but inevitable resignation.

Carnap went on to say that physics could, of course, explain the sequences of events seen in nature; but it was up to psychology to account for "the peculiarities of man's experiences with respect to time, including his different attitude towards past, present, and future." Einstein listened, and eventually told his friend that "there is something essential about the Now which is just outside the realm of science."

That is the voice (or at least the echo) of the doubting Einstein – the Einstein who presumably embraced all that his theory was saying yet perhaps wondered if it was still, somehow, incomplete. But it is another late-Einstein passage that writers and biographers prefer to quote: a trio of sentences that seem to find a place in just about every Einstein book. In these poignant words, we hear a very different man, one who – perhaps surprisingly – sounds far more confident in the bizarre world he had helped to create. The words come from a letter he wrote very near the end of his life. His close friend and former patent-office colleague, Michele Besso, had passed away. Einstein wrote a consoling letter to Besso's family; it is dated March 21, 1955 – less than a month before his own death.

"Now he has departed this strange world a little ahead of me," Einstein wrote of his friend's passing. "That signifies nothing. For us believing physicists, the distinction between past, present, and future is only a stubbornly persistent illusion."

NOTES

INTRODUCTION

1   **"If we are aware of anything . . ."** Lucas (1973), p. 8.

1   **"Time passes. Listen . . ."** Thomas, Dylan. *Under Milk Wood*. London: J.M. Dent & Sons, 1954. p. 3.

1   **"I've completely solved . . ."** quoted in Calaprice (2005), p. 216.

2   **"Even when it is dark . . ."** Aristotle, *Physics*, quoted in Lucas (1973), p. 12.

4   **"What, then, is time? . . ."** St. Augustine, *Confessions* (11:14), http://www.leaderu.com/cyber/books/augconfessions/bk11.html.

Chapter 1: HEAVENLY CLOCKWORK

9   **"The first grand discovery was time . . ."** Boorstin (1983), p. xvii.

10 **"Their tools were stone and wood . . ."** Clare Tuffy interview, May 2007.

11 **"As an astronomer . . ."** Tom Ray interview, May 2007.

11 **"There can be no doubt . . ."** Andrew B. Powell, "Newgrange – Science Or Symbolism," *Proceedings of the Prehistoric Society*, vol. 60 (1994), pp. 85-96, p. 86.

13 **"At exactly 8:54 hours GMT . . ."** Quoted in Ruggles (1999), p. 17.

16 **". . . probably had a rudimentary conception of time . . ."** John Shea interview, November 2003.

17 **"There could be no more compelling . . ."** Klein (2002), p. 189.

19 **". . . first appearance of religious ideologies."** Mithen (1996), p. 174.

19 **"these advantages were paid for . . ."** Fraser (1987), p. 14.

20 **"I believe that what we have . . ."** Anthony Aveni interview, May 1999.

21 **"Fitting an artifact . . ."** Aveni (1995b), p. 70.

23 **". . . one of the most notorious examples . . ."** Clive Ruggles, "Astronomy and Stonehenge," *Proceedings of the British Academy*, vol. 92 (1997), pp. 203-229, p. 203.

24 **"Statistically, the odds are in favour . . ."** Burl (1976), p. 53.

24 **If the builders were thinking astronomically . . .** Joshua Pollard and Clive Ruggles, "Shifting Perspectives: Spatial Order, Cosmology, and Pattern of Depositions At at Stonehenge," *Cambridge Archaeological Journal*, vol. 11, no. 1 (2001), pp. 69-90, p. 71.

25　**"But even if no eclipse ..."** Aveni (1995a), p. 25.

25　**"I am convinced ..."** Aveni (1995a), p. 31.

25　**No wonder that, in the Middle Ages ...** Colin Renfrew, "Setting the Scene: Stonehenge in the Round," *Proceedings of the British Academy*, vol. 92 (1994), pp. 3-14, p. 4.

25　**"... the cremated remains ..."** Philip Jackman, "A mystery solved?" *The Globe and Mail,* May 30, 2008, p. A2.

26　**"... a place of social gathering ..."** Aveni (1995a), pp. 26-27.

26　**"a reference to the past ..."** Pollard and Ruggles, p. 80.

26　**"Stonehenge always embodied ..."** ibid., p. 87.

26　**"... timeless frame of reference ..."** Alasdair Whittle, *Proceedings of the British Academy,* vol. 92 (1994), pp. 145-66, p. 163.

27　**"... the earliest genuine depiction ..."** quoted in Tony Paterson, "Gold star chart points way to German 'Stonehenge,'" *The Daily Telegraph,* Oct. 6, 2002 (online edition).

27　**... the Towers of Chankillo** Ivan Ghezzi and Clive Ruggles, "Chankillo: A 2,300-Year-Old Solar Observatory In in Coastal Peru," *Science,* vol. 315, no. 5816 (March 2, 2007), pp. 1239-43.

Chapter 2: YEARS, MONTHS, DAYS

33　**"This year has an additional ..."** quoted in Robert Hannah, "The moon, the sun, and the stars," in McCready (2001), p. 59.

34　**Yet centuries would pass ...** Duncan (1998), p. 17.

35　**"... to increase or decrease taxes ..."** ibid., p. 30.

35　**"... a certain rule to make ..."** Plutarch, *Lives,* (ed. Charles E. Eliot). Danbury, Conn.: Grolier Enterprises, 1980, p. 311.

36　**"the last year of confusion."** Duncan (1998), p. 33.

36　**"had been thought of as a ..."** ibid., p. 37.

37　**... a "suburb" of Machu Picchu ...** Thomas H. Maugh II, "Lost Incan 'suburb' in Andes rediscovered," *The Boston Globe,* Nov. 9, 2003 (online edition).

38　**"... overly precise ..."** Duncan (1998), p. 138.

39　**"the Maya divine temporal ..."** Aveni (1995a), p. 108.

39　**"... apples with oranges"** Duncan (2000), p. 186.

40　**"the king came to be ..."** David Stuart, "Kings of Stone," *Res: Anthropology and Aesthetics,* Spring/Autumn 1996, pp. 149-72, pp. 165-66.

40　**"The Maya were fatalists ..."** Aveni (1995a), p. 102.

41　**"This is the guy ..."** David Stuart interview, September 2003.

44　**The Beatles song ...** Steel (2000), p. 73.

46　**"invented a reason to disagree with ..."** Steel (2000), p. 98.

47 **... between the Catholic Curch and science** see, for example, John L. Heilbron, *The Sun in the Church: Cathedrals as Solar Observatories.* Cambridge, Mass.: Harvard University Press, 1999.

50 **"... a Trojan horse ..."** quoted in Duncan (1998), p. 213.

50 **The bill passed ...** ibid., p. 225.

50 **Again, ordinary people balked ...** ibid., p. 228.

Chapter 3: HOURS, MINUTES, SECONDS

52 **"... hung a great silver chain ..."** Jonathan Swift, *Gulliver's Travels,* (ed. John Hayward). New York: Random House, 1939. p. 30.

53 **He's said to have enjoyed ...** Geoff Chester, "Lighthouse of the skies," *Astronomy,* Aug. 2007, pp. 58-63; www.usno.navy.mil.

53 **"... seduced by the timekeeping art ..."** Demetrios Matsakis interview, July 2007.

57 **... thirteen kinds of sundials...** Boorstin (1983), p. 28.

57 **"Confound him, too ..."** quoted in McCready (2001), p. 121; Boorstin (1983), p. 28.

58 **When European clocks arrived ...** Boorstin (1983), p. 61.

59 **"The tick-tock of the clock's ..."** ibid., p. 39.

59 **"Here was man's declaration ..."** ibid., p. 39.

60 **... was installed at Dunstable Priory ...** Dale (1992), p. 20.

60 **"There is no doubt in my mind ..."** Bryson, Bill, *Notes from a Small Island.* New York: Harper Collins, 1995 (2001 ed.), p. 86.

61 **"Of course a clock working for ..."** John Plaister interview, August 1999.

62 **"We always say he should have ..."** Frances Neale interview, August 1999.

63 **It even had a tiny alarm ...** Landes (1983), p. 87.

63 **"was more a consequence ..."** Sara Schechner, "The time of day," in McCready (2001), pp. 121-139, p. 134.

64 **... historian Alfred Crosby ...** Crosby's idea is summarized in Anthony Aveni, "Time's Empire," *Wilson Quarterly,* vol. 23, no. 3 (1998), pp. 44-57.

64 **"In a relatively brief span of years ..."** Anthony Aveni, "Time's Empire," p. 47.

64 **"To redeem time is to see that ..."** from Baxter's "Christian Directory" of 1664, quoted in Whitrow (1988), p. 160.

65 **It may not be a coincidence that ...** Boorstin (1983), p. 72; McCready (2001), p. 166.

66 **Charles II appointed John Flamsteed ...** Eric G. Forbes, *Greenwich Observatory,* vol. 1. London: Taylor & Francis, 1975. p. 19-20.

68 **"Poor old Harrison spent ..."** Jonathan Betts interview, August 1999.

70 **"Steam power was the driving force ..."** Whitrow (1988), p. 160.

70 **"The clock, not the steam engine ..."** quoted in Whitrow (1988), p. 164.

71 **"What changes may now occur ..."** quoted in Blaise (2000), p. 140.

72 **"London time is kept at all stations ..."** quoted in Thelma C. Landon, "The Father of Standard Time," *Canadian Geographic,* Feb./Mar. 1990, pp. 74-81, p. 76.

73 **... French time was officially ...** Blaise (2000), p. 206.

73 **... was a Western creation ...** See Nicholas Hune-Brown, "Timing is Everything," *The Toronto Star,* Nov. 4, 2007, p. D1, D9.

74 **"Sunrise in Atlanta is at a ..."** Michael O'Malley interview, November 1999.

75 **The best of them could keep time to ...** Dale (1992), p. 60.

76 **... atomic fountain clocks ...** See, for example, Quinn Norton, "How Super-Precise Atomic Clocks Will Change the World in a Decade," *Wired* (online), Dec. 12, 2007.

76 **Researchers at the University of Tokyo ...** Paul Marks, "The Most Accurate Clock of All Time," *New Scientist,* May 18, 2005 (online edition).

76 **... In fact, the day is getting longer ...** Dale (1992), p. 61.

77 **Without these corrections ...** Michelle Stacey, "Clash of the Time Lords," *Harper's Magazine,* Dec. 2006, pp. 46-56, p. 50.

77 **The international body that decides ...** Michelle Stacey (2006), p. 56.

Chapter 4: IN TIME'S GRASP

79 **"In view of all you have to do ..."** quoted in Whitrow (1988), pp. 110-11.

80 **"... prisoners of the present ..."** quoted in Carl Honoré, "Slowing the world," *The National Post,* Jan. 26, 2002, pp. B1, B6.

80 **"There isn't enough time ..."** quoted in Alexandra Gill, "Sleep no more," *The Globe and Mail,* April 1, 2006, p. F9.

80 **Chaucer had no notion ...** Macey (1994), p. 443.

80 *nunc et in hora mortis nostrae* The text of the prayer has evolved over the centuries; the modern version, quoted here, dates from the mid-16th sixteenth century.

82 **"... woven together as if in ..."** David Pankenier interview, September 2003.

82 **"temporal harmony within the person ..."** Fraser (1987), p. 19.

82 **In the Hindu faith, cyclic time ...** Gorst (2001), p. 4; also John Bowker (ed.) (ed.), *The Oxford Dictionary of World Religions.* Oxford: Oxford University Press, 1997, p. 980.

83 **"One plunges into time's ..."** Philip Novak, "Buddhist Meditation and the Consciousness of Time," *Journal of Consciousness Studies,* vol. 3, no. 3 (1996), pp. 267-277, p. 277.

83-84 **"The temporal logic seems to be ..."** Aveni (1995b), p. 171.

84 **"... he will probably tell you ..."** Aveni (1995a), p. 93.

85   **"[For] these frequently mobile ..."** Aveni (1995a), p. 93; see also. Gell (1992), pp. 300-305.

85   **"In the Umeda 'week' ..."** Gell (1992), p. 88.

85   **"... the moon is like a tuber ..."** Gell (1992), p. 291.

86   **"The role played by the first wife ..."** Chap Kusimba interview, August 2003.

86   **"a long *past,* a *present* ..."** John Mbiti, *African Religions and Philosophy.* New York: Praeger Publishers, 1969, p. 17.

86   **"*Actual time* is therefore ..."** ibid., p. 17.

86   **Mbiti's views have received ...** John A.A. Ayoade, "Time in Yoruba Thought," in Richard A. Wright (ed.), *African Philosophy: An Introduction.* Washington: University Press of America, 1979, p. 95.

87   **"There is no concept of time ..."** Pritchard (1997), p. 11.

87   **"... contains no words ..."** This is Whitrow's summary of Whorf's conclusions, in Whitrow (1988), p. 8.

87   **"Hopi is not a timeless ..."** Gell (1992), p. 127.

87   **"... have successfully developed ..."** Whitrow (1988), p. 9.

88   **... the Aymara point forward ...** Rafael Núñez and Eve Sweetser, "With the Future Behind Them: Convergent Evidence From Aymara Language and Gesture in the Crosslinguistic Comparison of Spatial Construals of Time," *Cognitive Science,* vol. 30 (2006), pp. 401-450.

89   **"Aboriginal concepts of time ..."** Howard Morphy, "Australian Aboriginal Concepts of Time," in Lippincott (2000), p. 267.

89   **"... Jerusalem in about the year A.D. 29 ..."** Brandon (1965), p. 29.

89   **"time, place and people were ..."** Mike Donaldson, "The End of Time? Aboriginal temporality and the British invasion of Australia," *Time & Society,* vol. 5, no. 2 (1996), pp. 187-207, p. 193.

90   **"There is no fairyland ..."** Gell (1992), p. 315.

91   **"They define each other ..."** quoted in Danielson (2000), p. 38.

92   **"Socrates and Plato and each ..."** quoted in Coveny and Highfield (1990), p. 26; Whitrow (1988), p. 43.

93   **"One might wonder whether ..."** quoted in Barnes (1997), p. 88.

93   **"there is a circle ..."** quoted in Coveney and Highfield (1990), p. 25.

93   **"We must say that the same ..."** quoted in Whitrow (1988), p. 46.

94   **"just as in this age ..."** quoted in Gorst (2001), p. 7.

94   **"... dissociated time from human events ..."** quoted in Whitrow (1988), p. 127.

95   **"clock and calendar time"** John Postill, "Clock and Calendar Time: A missing anthropological problem," *Time & Society,* vol. 11 11, no. 2/3 (2002), pp. 251-270, p. 251.

95   **"'chronoclasm' – the intentional destruction ..."** ibid., p. 252.

96   **"altered everyday work ..."** ibid., p. 255.

96   **"While other countries . . ."** Nishimoto Ikuko, "The 'Civilization' of Time: Japan and the adoption of the western time system," *Time & Society,* vol. 63, no. 2/3 (1997), pp. 237-259, p. 239.

96   **In 1873, a textbook . . .** ibid., p. 250.

96   **"anywhere else in the world . . ."** quoted in Brigitte Steger, "Timing Daily Life in Japan," *Time and Society,* vol. 15, no. 2/3 (2006), pp. 171-175, p. 171.

97   **Levine does not probe this . . .** Levine (1997), p. 10.

97   **". . . the train left late . . ."** ibid., p. 6.

98   **"When we attribute . . ."** ibid., p. 203.

98   **"American Indians like to . . ."** ibid., p. 10.

98   **. . . in Spanish, the same verb . . .** ibid., pp. 94-5.

98   **While it is apparently okay . . .** "Inuit artist accuses CRA staff of writing racist tax memo," www.cbc.ca, Oct. 26, 2007.

98   **The Kapauka people . . .** Levine (1997), p. 14.

99   **"Our century, which began . . ."** quoted in Wendy Parkins, "Out of Time: Fast subjects and slow living," *Time & Society,* vol. 13, no. 2/3 (2004), pp. 363-382, p. 372.

99   **"Have I gone completely . . ."** quoted in Zsuzsi Gartner, "What's your big hurry," *The Globe and Mail,* May 15, 2004, p. D6.

100   **"a bubble in which . . ."** Kate Zernicke, "Calling In in Late," *The New York Times,* Oct. 26, 2003, Section 9, p. 1, p. 11.

Chapter 5: THE PERSISTENCE OF MEMORY

101   **"Memory's vices are also . . ."** Schacter (1996), p. 206.

101   **"To think . . ."** quoted in Joshua Foer, "Remember This," National Geographic, Nov. 2007, p. 54.

101   **"I would query by what . . ."** quoted inWhitrow (1972), p. 28.

102   **"the astonishing hypothesis"** Francis Crick, *The Astonishing Hypothesis,* New York: Macmillan, 1994.

103   **"Many complex human behaviors . . ."** "How Does Your Brain Tell Time?" Press release from the University of California in Los Angeles, Jan. 29, 2007; see also David M. Eagleman et al, "Time and the Brain: How Subjective Time Relates to Neural Time," *The Journal of Neuroscience,* vol. 25, no. 45 (Nov. 9, 2005), pp. 10369-10371.

103   **"How internal clocks . . ."** Suddendorf and Corballis, "The evolution of foresight: What is mental time travel and is it unique to humans?" *Behavioral and Brain Sciences,* in press (2007).

103   **"suprachiasmatic nucleus"** Ralph Mistlberger, "Keeping time with nature," in McCready (2001), p. 33.

103   **"contributing to an emerging picture . . ."** David Eagleman (2005), p. 10369.

104   **"Despite its importance . . ."** ibid.

104   **"Considering this, I say, . . ."** quoted in Whitrow (1972), p. 28.

104   **". . . the workings of memory . . ."** Several recent books by distinguished scientists do attempt such an overview. Particularly noteworthy are neuroscientist Eric Kandel's *In Search of Memory* (2006) and psychologist Daniel Schacter's *Searching for Memory* (1996).

106   **"to wonder about things like . . ."** quoted in Barbara Turnbull, "Mastering the mind," *The Toronto Star,* Sept. 16, 2006 (online edition).

106   **"Most forms of memory . . ."** Endel Tulving, lecture at the University of Toronto, Sept. 25, 2007.

106   **"Remembering, for the rememberer . . ."** quoted in Schacter (1996), p. 17.

107   **"provides increased behavioural . . ."** Suddendorf and Corballis (2007).

107   **"What is the benefit of knowing . . ."** Tulving lecture (2007).

107   **Brain imaging studies have shown that . . .** Daniel L. Schacter et al, "Remembering the past to imagine the future: the prospective brain," *Nature Reviews – Neuroscience,* vol. 8 (Sept. 2007), pp. 657-661; for a popular account, see Jessica Marshall, "Future recall," *New Scientist,* 24 March 2007, pp. 36-40.

107   **"as a fundamentally prospective . . ."** Daniel Schacter (2007), p. 660.

107   **"We tend to think of . . ."** Daniel Schacter interview, May 2007.

108   **". . . part of a more general toolbox . . ."** Suddendorf and Corballis (2007); see also Schacter (1996).

109   **"completely rooted in the present . . ."** William A. Roberts, "Are Animals Stuck in Time?" *Psychological Bulletin,* vol. 128, no. 3 (2002) pp. 473-489, p. 473.

109   **". . . a shell of a person . . ."** Schacter interview (2007); see also Barbara Turnbull (2006).

109   **"Without memory, E. P. . . ."** Joshua Foer, "Remember This," *National Geographic,* Nov. 2007, pp. 32-56, p. 37, p. 55.

110   **". . . stuck in the present . . ."** This idea has been expressed by Endel Tulving, Sue Savage-Rumbaugh, Merlin Donald, and others; for an overview, see Thomas Suddendorf and Michael C. Corballis, "Mental Time Travel and the Evolution of the Human Mind," *Genetic, Social, and General Psychology Monographs,* Vol,ol. 123, no. 2 (1997), pp. 133-167.

110   **"Remembering past events . . ."** quoted in William A. Roberts (2002), p. 473.

110   **"'linguistic outputs' of trained . . ."** Suddendorf and Corballis, 2007.

111   **"But what are they thinking about?"** quoted in Eric Jaffe, "Mental Leap: "What apes can teach us about the human mind," *Science News,* vol. 170, no. 10 (Sept. 2, 2006), online edition.

111   **the birds will recover ...** Originally published in *Nature;* these results
       are summarized in Nicola S. Clayton et al, "Can animals recall the past
       and plan for the future?" *Nature Reviews – Neuroscience,* vol. 4 (Aug. 2003),
       pp. 685-691.

112   **Because the birds were not ...** C.R. Raby et al, "Planning for the future by
       western scrub-jays," *Nature,* vol. 445 (22 Feb. 2007), pp. 919-921; also dis-
       cussed in Suddendorf and Corballis (2007); Carl Zimmer, "Time in the
       Animal Mind," *The New York Times,* April 3, 2007 (online edition).

112   **"can spontaneously plan for tomorrow ..."** C.R. Raby et al (2007), p. 919.

112   **"can anticipate and plan for ..."** William A. Roberts, "Mental Time Travel:
       Animals Anticipate the Future," *Current Biology,* vol. 17, no. 11 (2007), pp.
       418-420, p. 418.

112   **"may be aware of only ..."** William A. Roberts (2002), p. 486.

112   **"the ability to manage past ..."** Thomas R. Zentall, "Mental time travel in
       animals: A challenging question," *Behavioural Processes,* vol. 72 (2006),
       pp. 173-183, p. 173.

112   **"unconvinced that any of these ..."** Suddendorf and Corballis (2007).

112   **Suddendorf and Corballis insist ...** Suddendorf and Corballis (2007).

113   **He says he is glad ...** Tulving lecture (2007).

113   **"It emerged more recently ..."** Tulving lecture (2007).

113   **"To entertain a future event ..."** Suddendorf and Corballis (2007).

114   **It is interesting to note that ...** Whitrow (1988), pp. 5-6.

114   **Another ability that may be ...** This has been investigated by
       H.M. Wellman, J. Perner, and others, and is summarized in Suddendorf
       and Corballis (2007).

114   **"... the ultimate step ..."** ibid.

114   **Not that every realization ...** ibid.

115   **"... because of continued ..."** ibid.

115   **"The mental reconstruction of past ..."** ibid.

115   **Although young children can ...** Charles Nelson, "Ask Discover: Why don't
       we remember things from when we were babies?" *Discover* Feb. 2005, p. 13;
       Jamie Baker, "Why early memories disappear," *The National Post,* Sept. 29,
       2005 (online edition).

116   **... between the ages of three and five ...** Janie Busby and Thomas
       Suddendorf, "Recalling yesterday and predicting tomorrow," *Cognitive
       Development,* vol. 20 (2005), pp. 362-372.

116   **As William Roberts suggests ...** William A. Roberts (2002), p. 473.

116   **This is also the age ...** These results are summarized in Suddendorf and
       Corballis (2007).

116   **Such experiments are fraught ...** Busby and Suddendorf (2005), p. 370.

116 **Interestingly, most ten-year-olds ...** Whitrow (1988), p. 6.

117 **"difficult to relate ..."** ibid., p. 7.

117 **"We have no dedicated sense ..."** Gell (1992), p. 92.

118 **It is hardly a surprise ...** An excellent recent book on the subject is Daniel Schacter's *The Seven Sins of Memory* (2001).

118 **... they also frequently ...** Daniel L. Schacter, "The Cognitive Neuroscience of Constructed Memory: Remembering the Past and Imagining the Future," *Philosophical Transactions of the Royal Society (B),* in press; Daniel Schacter (1996), p. 103.

118 **"The positive spin on this ..."** Schacter interview; see also Daniel L. Schacter and Donna Rose Addis, "The ghosts of past and future," *Nature,* vol. 445 (4 Jan. 2007), p. 27; Daniel Schacter (2001), Chapter 8.

119 **"billions and billions"** Carl Sagan, *Billions and Billions.* New York: Ballantine Books, 1998, pp. 3-4.

119 **Prior interviews with close relatives ...** Elizabeth F. Loftus, "Creating False Memories," *Scientific American,* Sept. 1997, pp. 70-75.

119 **... other controversial practices ...** See, for example, Elizabeth Loftus and Katherine Ketcham, *The Myth of Repressed Memory*, New York: St. Martin's Press, 1994.

120 **"flashbulb memories"** Schacter (1996), pp. 195-201.

120 **... incredibly, a survey by ...** "30% of Americans cannot say what year 9/11 attacks happened, poll finds," *The National Post,* Aug. 10, 2006, p. A18.

120 **... a detailed study of 9/11 memories ...** Elizabeth Phelps interview, July 2007; also Tali Sharot et al, "How personal experience modulates the neural circuitry of memories of September 11," *Proceedings of the National Academy of Sciences,* vol. 104, no. 1 (Jan. 2, 2007), pp. 389-394.

122 **"... after the *Challenger* disaster ..."** Daniel Greenberg, "Flashbulb memories: How psychological research shows that our most powerful memories may be untrustworthy," *Skeptic,* vol. 11, no. 3 (Winter 2005), accessed through InfoTrac.

122 **"It's hard to convince people ..."** Phelps interview (2007).

123 **As UCLA psychologist ...** My account is based on Daniel Greenberg (2005).

123 **"I was in Florida ..."** CNN, quoted in Daniel Greenberg (2005).

123 **"Bush remembers senior adviser ..."** *The Washington Post,* quoted in Daniel Greenberg (2005).

124 **"I was sitting there ..."** White House press release, quoted in Daniel Greenberg (2005).

124 **In one Dutch study ...** Daniel Greenberg (2005).

125 **"the President, like most Americans ..."** Daniel Greenberg (2005).

Chapter 6: ISAAC'S TIME

126 **"Nature, and Nature's Laws . . ."** quoted in Coveney and Highfield (1990), p. 39.

126 **"No closer to the gods . . ."** Cohen and Whitman (1999), p. 380.

126 **". . . sober, silent, thinking lad"** quoted in Westfall (1994), p. 13.

127 **"In those days, I was in . . ."** ibid., p. 39.

127 **". . . to decline correspondencies . . ."** ibid., p. 109.

128 **"write on his Desk . . ."** ibid., p. 162.

128 **"Because mathematicians frequently . . ."** quoted in Coveney and Highfield (1990), p. 29.

129 **"Absolute, true, and mathematical . . ."** Cohen and Whitman (1999), p. 408.

129 **"Time does not imply motion. . ."** quoted in Whitrow (1988), p. 128.

130 **"a mode of thinking"** quoted in Turetzky (1998), p. 71.

130 **"the analogy of time with space"** Turetzky (1998), p. 72.

130 **"Absolute space . . ."** Cohen and Whitman (1999), p. 408.

132 **"is a fantasy that blinds . . ."** Lee Smolin, "What Is Time?" in John Brockman (1995), p. 236.

133 **". . . is plainly maintaining that God . . ."** quoted in Keith Ballard, "Leibniz's Theory of Space and Time," *Journal of the History of Ideas,* vol. 21, pp. 49-65, p. 53.

133 **"For how can a thing . . ."** Alexander (1956), pp. 72-73.

133 **The debate continues . . .** J.R. Lucas, "Time and Religion," in Ridderbos (ed.) (2002), pp. 143-167, p. 162.

133 **Newton's voluminous theological . . .** see, for example, Westfall (1994).

134 **"He is eternal and infinite . . ."** Cohen and Whitman (1999), p. 941.

134 **"This God of dominion . . ."** Stephen Snobelen, "'The true frame of nature': Isaac Newton, Heresy, and the Reformation of Natural Philosophy," in John Brooke and Ian Maclean (eds.), *Heterodoxy in Early Modern Science and Religion.* Oxford: Oxford University Press, 2005, p. 254.

134 **"This most beautiful system . . ."** Cohen and Whitman (1999), pp. 940-941. Newton also makes a similar argument about the biological world.

134 **. . . the clockwork metaphor . . .** Thanks to James Robert Brown for helpful comments on this matter.

135 **"If time flows . . ."** quoted in Lockwood (2005), p. 13.

135 **. . . Huw Price points out . . .** Price (1996), p. 13.

136 **"We may regard the present state . . ."** "Ask Science," *The New York Times,* March 17, 2006 (online edition).

137 **"If the motion of every . . ."** Thomson, William, "Kinetic Theory of the Dissipation of Energy," *Nature,* vol. 232 no. 9 (1874), p. 442.

139   **"the supreme position ..."** quoted in Savitt (1995), p. 1.

141   **"We have looked through the ..."** ibid., p. 1.

143   **"... without creation or destruction ..."** quoted in Turetzky (1998), p. 10.

149   **"... quarreling over a toy ..."** Julian Barbour interview, May 2007.

151   **"a philosophical health warning"** Simon Saunders, "Clock Watcher," *The New York Times*, March 26, 2000 (online edition).

151   **"There goes the man that ..."** quoted in Westfall (1994), p. 190.

Chapter 7: ALBERT'S TIME

152   **"Relativity has taught us ..."** quoted in Coveney and Highfield (1990), p. 70.

152   **"I see the past ..."** quoted in Pickover (1998), p. 6.

153   **"Einstein was so happy ..."** Ruth Aegler interview, June 2004.

156   **"You will discover not the least ..."** Galileo (1967), pp. 186-7.

157   **"There seems to be no such ..."** quoted in Stachel (1998), p. xxxix.

157   **"know, or be able to establish ..."** ibid., p. xxxix.

157   **Peter Galison has argued ...** Galison (2003).

158   **"The first thing we'll do ..."** quoted in Isaacson (2007), p. 46.

158   **"cannot be absolutely defined ..."** quoted in Calaprice (2005), p. 216.

159   **"for several valuable suggestions"** Albert Einstein, "On the Electrodynamics of Moving Bodies," in Stachel (1998), p. 159.

163   **"He has no stakes at all ..."** Gerald Holton interview, November 2004.

166   **"one of the deepest insights ..."** Greene (1999), p. 36.

166   **"There is no audible tick-tock ..."** quoted in Isaacson (2007), p. 128.

166   **"... I don't have a watch."** quoted in Fölsing (1997), p. 266.

167   **In Gwinner's experiment ...** Elizabeth Quill, "Time Slows When You're on the Fly," ScienceNOW website, http://sciencenow.sciencemag.org/cgi/content/full/2007/1113/2.

167   **"Henceforth, space on its own ..."** quoted in Fölsing (1997), p. 189.

168   **... may yet be in your future ...** I have omitted the physics behind this assertion, but the interested reader may turn to Roger Penrose's *The Emperor's New Mind* or Michael Lockwood's *The Labyrinth of Time* for a more detailed discussion. Penrose (pp. 260-61) gives a remarkable example involving a space fleet from the Andromeda Galaxy, some two million light years away, intent on destroying Earth. If you and I pass each other on the street – even at normal walking speeds – we can disagree about what time it is on Andromeda by several days. For one of us, the fleet is already on its way; for the other, the decision to launch has not even been made!

170   **"The primitive, subjective feeling ..."** Einstein and Infeld (1938), p. 180.

171   **"Let's say that an event ..."** I've taken this example from Michael Lockwood's version of Putnam's argument, outlined in *The Labyrinth of Time* (2005).

171   **"To take the spacetime view ..."** ibid., p. 68-69.

171   **"The very division ..."** Davies (1995), p. 71.

171   **"... made a deep impression ..."** Albert Einstein, "How I Created the Theory of Relativity" (trans. Yoshimasa Ono). *Physics Today*, August 1982, p. 47.

172   **"In all my life ..."** quoted in Pais (1982), p. 216.

173   *The Times* (London), 7 November 1919, p. 12; *The New York Times*, 10 November 1919, p. 17.

174   **... were able to disentangle ...** Lockwood (2005), p. 80.

174   **... a U.S. lab in Boulder, Colorado ...** Coveney and Highfield (1990), p. 95.

176   **"... services to theoretical physics ..."** http://nobelprize.org; Isaacson (2007), p. 314.

178   **"In the act of measurement ..."** Paul Davies, "That Mysterious Flow," *Scientific American,* Sept. 2002, pp. 40-47, p. 47.

178   **"Anyone who is not ..."** quoted in Gribbin (1984), p. 5.

178   **"Newton, forgive me ..."** Einstein (1979), p. 31.

Chapter 8: BACK TO THE FUTURE

180   **"'I know,' he said ..."** H.G. Wells, *The Time Machine: An Invention.* New York: Random House, 1931. p. 82.

180   **After the war ...** Mallett (2006), pp. 2-3.

180   **"completely crushed"** Ronald Mallett interview, May 2007.

181   **"Black holes are what people ..."** Mallett interview (2007).

181   **Mallett remained in the ...** Mallett interview (2007); also Michael Brooks, "Time Twister," *New Scientist,* May 19, 2001.

182   **... in peer-reviewed physics journals** Ronald Mallett, "Weak gravitational field of the electromagnetic radiation in a ring laser," *Physics Letters A,* vol. 269 (2000), p. 214; Ronald Mallett, "The Gravitational Field of a Circulating Light Beam," *Foundations of Physics,* Vol. 33, No. 9 (Sept. 2003), pp. 1307-1314.

182   **"... a distant improbability ..."** quoted in Michael Brooks (2001), p. 19.

182   **"greater than the radius ..."** Ken D. Olum and Allen Everett, "Can a circulating beam of light produce a time machine?", *Foundations of Physics Letters,* vol .18, p. 379-385 (Oct. 2004), p. 379.

184   **"In principle – if quantum ..."** John Cramer interview, April 2007.

184   **"I feel a little uncomfortable ..."** Cramer interview (2007).

185 **"... overturn our most cherished notions ..."** Patrick Barry, "What's done is done," *New Scientist,* Sept. 30, 2006, pp. 36-39, p. 36.

186 **"... a crucial breakthrough ..."** Clute and Nicholls (1995), p. 1225.

186 **... a flood of time travel stories ...** A wonderfully comprehensive account of such stories can be found in Paul Nahin's *Time Machines: Time Travel in Physics, Metaphysics, and Science Fiction* (1999).

187 **"the main work to consult"** Douglas Adams, *The Restaurant at the End of the Universe.* London: Pan Books, 1980 (1983 ed.), pp. 79-80.

188 **So far he as aged ...** Dennis Overbye, "A Trip Forward in Time. Your Travel Agent: Einstein," *The New York Times,* June 28, 2005, p. F4; www.wikipedia.org, "Sergei Krikalev."

188 **Let's say you want to circumnavigate ...** The example and the calculations are from Michael Lockwood's *The Labyrinth of Time* (2005), p. 48.

189 **"There is every reason to believe ..."** Greene (2004), p. 449.

189 **"Einstein's equations of general relativity ..."** Krauss (1995), p. 15.

191 **It didn't take long for wormholes ...** Thorne (1994), pp. 483-484.

191 **"the idea of time machines ..."** Toomey (2007), p. 18.

192 **A variety of other equally esoteric ...** A concise roundup of possible time machine mechanisms can be found in Ivan Semeniuk, "No going back," *New Scientist,* Sept. 20, 2003, pp. 28-32; for a more detailed account, see Davies (2001) and Nahin (1999).

194 **... philosopher David Lewis ...** David Lewis, "The Paradoxes of Time Travel," *American Philosophical Quarterly,* vol. 13, no. 2 (April 1976), pp. 145-52.

195 **One way or another ...** This was essentially the solution proposed by Lewis in his 1976 paper.

196 **"without reference to what ..."** David Deutsch and Michael Lockwood, "The Quantum Physics of Time Travel," *Scientific American,* March 1999, pp. 68-74, p. 71.

196 **When any one *particular* ...** Lockwood (2005), p. 172. Lewis had expressed essentially the same idea, though formulated in less technical terms.

196 **"If you time-travel to the past ..."** Greene (2004), p. 454.

197 **"the only solutions ..."** quoted in Nahin (1999), p. 272. As Nahin points out, the principle had been set out earlier by the Russian philosopher Igor Novikov.

197 **"If a time traveler is going to ..."** Nicholas Smith, "Bananas Enough for Time Travel," *British Journal of the Philosophy of Science,* vol. 48, 1997, pp. 363-389, p. 366.

200 **"It is the explanation ..."** Deutsch (1997), p. 51.

201 **"*The laws of physics conspire ...*"** Hawking (2001), p. 153.

201 **"the best evidence we have that . . ."** Hawking (1994), p. 154. (Arthur C. Clarke made the same point more than twenty years earlier.)

201 **Another possibility is that we're . . .** Toomey (2007) suggests several more possible reasons for the absence of time-traveling tourists.

201 **"and there is no reason . . ."** David Deutsch and Michael Lockwood, "The Quantum Physics of Time Travel," *Scientific American,* March 1994, pp. 68-74, p. 74.

201 **"does begin to seem . . ."** Jonathan Leake and Rajeev Syal, "Hawking: we'll be able to travel back in time," *The Sunday Times,* Oct. 1, 1995, p. 1.

202 **"will need to find another . . ."** Leonard Susskind, "Wormholes and Time Travel? Not Likely," http://arxiv.org/abs/gr-qc/0503097v3 (April 8, 2005), p. 4.

202 **Some physicists have also suggested . . .** Ivan Semeniuk, "No going back," *New Scientist,* Sept. 20, 2003, pp. 28-32.

202 **"Even if it turns out . . ."** Hawking (2001), p. 147.

Chapter 9: IN THE BEGINNING

204 **"We aspire in vain . . ."** Charles Lyell, *Principles of Geology* (vol. 3). New York: Johnson Reprint Corp., 1969. p. 384.

205 **. . . millions of years of weathering . . .** "Canyon de Chelly," pamphlet published by the U.S. National Park Service, U.S. Department of the Interior.

205 **"Two thousand years . . ."** Gorst (2001), pp. 3-4.

207 **Later commentators usually . . .** ibid., pp. 34-39.

208 **"In a stroke, he had . . ."** ibid., p. 104.

209 **"Is it not that being . . ."** quoted in Gorst (2001), p. 119.

209 **. . . Newton could not accept . . .** Toulmin and Goodfield (1965), pp. 146-7.

209 **"The winds and water disintegrate . . ."** quoted in Toulmin and Goodfield (1965), p. 64.

210 **". . . no vestige of a beginning . . ."** quoted in Gorst (2001), p. 134.

210 **"Millions and whole myriads . . ."** quoted in Toulmin and Goodfield (1965), p. 133.

210 **"By 1750, men could contemplate . . ."** ibid.

211 **"The sound which, to the student . . ."** quoted in Gorst (2001), p. 146; Toulmin and Goodfield (1965), p. 170.

212 **"When seeing a thing . . ."** quoted in Gribbin (1999), p. 19.

213 **"must have succeeded each other . . ."** quoted in Gorst (2001), p. 167.

213 **". . . a *time* bomb"** Ferris (1988), p. 245.

214 **"The living plants and animals . . ."** Archibald Geike, "Geological Change," in Shapley (1943), pp. 112-3.

214 **"He who . . . does not admit . . ."** quoted in Ferris (1988), p. 245.

214 **"thus increases the possible limit . . ."** ibid., p. 249.

215 **"For a public used to dealing ..."** Gorst (2001), p. 204.

216 **"... not meant to be gaped at"** Arthur Eddington, "The Milky Way and Beyond," in Shapley (1943), p. 93.

217 **"scattered through space ..."** Edwin Hubble, "The Exploration of Space," in Ferris (1991), p. 336.

221 **astronomers now believe ...** For an excellent overview, see Wendy Freedman and Michael Turner, "Cosmology in the New Millennium," *Sky & Telescope,* Oct. 2003, pp. 30-41.

222 **... the first crucial paper ...** Alan Guth, "The Inflationary Universe: A Possible Solution to the Horizon and Flatness Problems," *Physical Review D,* vol. 23 (1981), pp. 347-56.

222 **"Conceivably, everything ..."** quoted in Danielson (2000), pp. 482-3.

223 **"... the observations certainly show ..."** Alan Guth interview, March 2003. For a recent popular account, see Adam Frank, "Seeing the Dawn of Time," *Astronomy,* Aug. 2005, pp. 34-39.

223 **"Many of the greatest minds ..."** Gorst (2001), p. 291.

224 **"The larger the universe ..."** Ferris (1997), p. 305.

Chapter 10: BEYOND THE BIG BANG

225 **"Our picture of physical reality ..."** Penrose (1989), p. 480.

228 **These brane-world models ...** Popular accounts include Gabriele Veneziano, "The Myth of the Beginning of Time," *Scientific American,* May 2004, pp. 54-65; Michael Lemonick, "Before the Big Bang," *Discover,* Feb. 2004, pp. 35-41; Paul Steinhardt, "A Cyclic Universe," SEED, July-Aug. 2007, pp. 32-34.

228 **"that God put fossils in the rocks ..."** Stephen Hawking lecture, U.C. Davis, March 23, 2003.

228 **"You should imagine ..."** Paul Steinhardt interview, March 2003.

229 **The cosmos as a whole ...** Andrei Linde, "The Self-Reproducing Inflationary Universe," *Scientific American,* Nov. 1994, pp. 48-55.

230 **"our universe is simply ..."** Edward P. Tyron, "Is the Universe a Vacuum Fluctuation?" *Nature,* vol. 246 (1973), pp. 396-7.

230 **"... tells us that fundamentally ..."** Lisa Randall, untitled essay, *New Scientist,* Nov. 18, 2006, p. 49.

231 **"clearly missing something very big"** Nima Arkani-Hamed lecture, hosted by the Perimeter Institute for Theoretical Physics, held at Waterloo Collegiate Institute, Waterloo, Ont., Feb. 7, 2007.

231 **They would ideally like ...** Greene (2004), pp. 489-481; James Glanz, "Physics' Big Puzzle Has Big Questions: What Is Time?" *The New York Times,* June 19, 2001.

232 **"But we've learned . . ."** David Gross interview, October 2003.

234 **Because of this connection . . .** See, for example, Price (1996), p. 51. Price says that most physicists believe that the thermodynamic arrow explains the radiative arrow; however, he is personally doubtful of this argument.

235 **"rather like throwing a ball . . ."** Davies (1995), p. 209.

235 **"likely to be of little . . ."** Greene (2004), p. 145. See also Davies (1995), pp. 208-213.

236 **. . . have spent years grappling . . .** For example, John Cramer (Chapter 8) suspects that the cosmological arrow is paramount, and that it causes the radiative arrow, which in turn causes the thermodynamic arrow (Cramer interview, 2007). Several recent books examine the arrow of time in detail, including Coveney and Highfield (1990), Savitt (1995), and Price (1996). There are also very good discussions in Hawking (1988), Penrose (1989), and Greene (2004).

236 **"polymath extraordinaire"** The quote is from Tim Folger, "If an Electron Can Be in 2 Places at Once, Why Can't You?" *Discover,* June 2005, pp. 28-35, p. 30.

236 **. . . a novel description of spacetime . . .** For an overview, see Roger Penrose, "Strings with a twist," *New Scientist,* 31 July 2004, pp. 26-29.

238 **"brain-aching"** The quote is from George Johnson of the *New York Times,* who mentions *The Emperor's New Mind* in his review of *The Road to Reality.* Of *Emperor,* he says: "Starting from scratch with Pythagoras and Plato, [Penrose] dismantles what is known about the nature of the universe and then puts it back together again." George Johnson, "A Really Long History of Time," *The New York Times,* Feb. 27, 2005, p. 14.

239 **"We *seem* to be moving . . ."** Penrose (1989), pp. 391-2.

239 **"According to relativity . . ."** Penrose (1994), p. 384.

239 **"Many of them are related . . ."** Roger Penrose interview, May 2007.

Chapter 11: ALL THINGS MUST PASS

245 **"Perceivest not . . ."** Lucretius, *On the Nature of Things,* Book 5.1, (trans. W.E. Leonard), in Fraser (1987), p. 33.

245 **"Eternity is very long . . ."** quoted in Rees (2001), p. 117.

246 **. . . oaks are typically harvested . . .** After being cited by Hillis in an essay posted on the organization's website, www.longnow.org, the story of the 500-year-old trees has circulated endlessly, even appearing in the *New York Times* and in the book *Deep Time* by Gregory Benford (1999). The warden of New College wrote in 2002: "No matter how often the story is denied, newspapers and radio journalists still insist on believing that [the

19th-century workers] used oak beams from trees that had been planted for the purpose almost five hundred years before. Since most structural oak was cut from trees of about a hundred and fifty years old, it would have been unlikely that anyone would plant it for use in five hundred years." (http://www.new.ox.ac.uk/pdfs/alumni_nc_news_nov2002.pdf).

246   **"I came to think of this ..."**
http://www.digitalsouls.com/2001/Brian_Eno_Big_Here.html.

248   **"It doesn't really look ..."** Alison Boyle interview, May 2007.

248   **"Hours are an arbitrary artifact ..."** quoted in Patricia Leigh Brown, "A Clock to See You Through the Next 10,000 Years," *The New York Times,* April 2, 2000, p. WK5.

249   **"the world's slowest computer"** Stewart Brand, *The Clock of the Long Now: Time and Responsibility – The Ideas Behind the World's Slowest Computer.* New York: Basic Books, 2000.

250   **... will peak at about 9 billion ...** Cocks (2003), p. 40.

251   **... a book called *L'An 2440* ...** Cornish (1977), p. 58; Clute and Nicholls (1995), p. 457.

252   **"Flight by machines ..."** www.brainyquote.com; http://www.nasa.gov/ centers/dryden/news/X-Press/stories/2004/013004/res_feathers.html.

252   **"There is no reason ..."** http://listverse.com/history/top-30-failed-technology-predictions/.

253   **"640,000 bytes of ..."** quoted in Kurzweil (2000), p. 170.

254   **"we'll be lucky ..."** Arthur C. Clarke, "2099 ... The Beginning of History," in Griffiths (1999), pp. 43-44.

254   **In an interview ...** The interview was with the German magazine *Focus,* and was later summarized in an article in the *Observer* (story by Nick Paton Walsh, Sept. 2, 2001, online edition).

254   **"any clear distinction ..."** Kurzweil (2000), p. x.

254   **"are on the cusp ..."** Kaku (1997), p. 5.

255   **"an escalating struggle ..."** Cocks (2003), p. 130.

255   **"A few adherents of ..."** Rees (2003), pp. 48-9.

255   **"There will always be ..."** ibid., p. 43.

257   **Gott came up with ...** The story is recounted in Gott (2002), pp. 207-9.

258   ***Homo sapiens* have been around ...** As Gott points out, similar arguments have been put forward by Australian physicist Brandon Carter and Canadian philosopher John Leslie. They used a different mathematical approach, based on a method known as "Bayesian statistics," but reached very similar conclusions. Leslie presents one version of the argument in his book *The End of the World* (1996).

258   **... and none of the dates ...** Richard Gott, personal communication.

258   **"an abstract mathematical model ..."** Freeman Dyson, "How Long Will the Human Species Last? An Argument with Robert Malthus and Richard Gott," in Brockman (1995), pp. 269-275, p. 271.

258   **"The knowledge of this ..."** ibid., p. 274.

259   **"... started taking bets ..."** Helen Carter, "Plenty of bets on Armageddon," *The Guardian,* Feb. 8, 1999 (online edition).

259   **"love with robots ..."** David Levy, *Love + Sex with Robots: The Evolution of Human-Robot Relationships.* New York: Harper Collins, 2007. The quotation appears in Robin Marants Henig, "Robo Love," *The New York Times,* Dec. 2, 2007, p. BK14.

260   **"replay the tape"** Stephen Jay Gould, *Wonderful Life: The Burgess Shale and the Nature of History.* New York: W.W. Norton and Company, 1989.

261   **"... a hospitable place."** Fred Adams interview, June 1998.

261   **... a bloated 168 million kilometers.** These are newer figures from Klaus-Peter Schröder and his colleagues at the University of Sussex, as quoted in Gribbin (2006), p. 250.

261   **"... the end of life on Earth."** Fred Adams, "Long Term Astrophysical Processes," in Bostrom and Cirkovic (in press).

262   **"Earth is thus evaporated ..."** Fred Adams, "Long Term Astrophysical Processes."

264   **"Since the expansion ..."** Fred Adams, personal communication.

266   **"The universe would be ..."** Fred Adams interview (1998).

267   **"... cloaked behind ..."** Fred Adams, "Long Term Astrophysical Processes."

268   **"The universe will look static ..."** quoted in J.R. Minkel, "A.D. 100 Billion: Big Bang Goes Bye-Bye," *Scientific American* (online edition), May 28, 2007.

268   **"All the labours ..."** Bertrand Russell, *A Free Man's Worship.* London: George Allen & Unwin Ltd., 1976. p. 10.

269   **... any living creatures ...** Interestingly, Krauss and Starkman suggest that copying our minds onto non-living material is the least of our problems. "While futuristic, the idea of shedding our bodies presents no fundamental difficulties ... Most philosophers and cognitive scientists regard conscious thought as a process that a computer could perform ... We still have many billions of years to design new physical incarnations to which we will someday transfer our conscious selves." Lawrence Krauss and Glenn Starkman, "The Fate of Life in the Universe," *Scientific American,* November 1999, pp. 58-65, pp. 62-3.

270   **"If they had made it ..."** quoted in Patricia Leigh Brown, "A Clock to See You Through the Next 10,000 Years," *The New York Times,* April 2, 2000, p. WK5.

270   "... the values of our own age ..." Brian Hayes, "Clock of Ages," *The Sciences,* Nov./Dec. 1999, pp. 9-13, p. 13.

Chapter 12: ILLUSION AND REALITY

272   **"Time is a river ..."** Jorge Luis Borges, "A New Refutation of Time," in *Labyrinths* (ed. Donald A. Yates and James E. Irby). New York: New Directions Publishing Corp., 1964. p. 234.

273   **"In essence, the self ..."** Patricia Churchland, "Do We Have Free Will?" *New Scientist,* Nov. 18, 2006, pp. 42-45, pp. 44-45.

274   **"There is nothing that corresponds ..."** Paul Davies interview, September 2007

274   **"It is in you ..."** Augustine, *Confessions* 11:27, in Fraser (1987), p. 34.

275   **"Time is not something objective ..."** quoted in Coveney and Highfield (1990), p. 28; Fraser (1987), p. 42.

275   **"This illusion cries out ..."** Paul Davies, "That Mysterious Flow," *Scientific American,* Sept. 2002, pp. 40-47, p. 47.

275   **"Our perception of ..."** Robert Jaffe interview, April 2007.

276   **"Our powerful sense ..."** quoted in Marcus Chown, "Clock-watchers," *New Scientist,* 1 May 2004, pp. 34-37, p. 34.

276   **"Something analogous to ..."** ibid., p. 35.

276   **"One implication ..."** ibid., p. 37.

277   **"a deeply innate ..."** Jaffe interview (2007).

278   **"Whenever I sit ..."** Greene (2004), p. 471.

279   **"Even when it is ..."** Aristotle, *Physics,* quoted in Lucas (1973), p. 12.

280   **Dummett gets the ball rolling ...** Michael Dummett, "Is Time a Continuum of Instants," *Philosophy,* vol. 75 (2000), pp. 497-515.

280   **"It is unclear how ..."** Rupert Read, "Is 'What is Time?' a Good Question to Ask?", *Philosophy,* vol. 77 (2002), pp. 193-209, p. 193.

280   **"appears not to understand"** Michael Dummett, "How should we conceive of time?", *Philosophy,* Vol. 78 (2003), pp. 387-396, p. 388.

280   **"... appears to go berserk ..."** ibid., p. 390.

280   **"does not take seriously *enough* ..."** Rupert Read, "Time to stop trying to provide an account of time," *Philosophy,* Vol. 78 (2003), pp. 397-408, p. 399.

280   **"... will never second-guess physics ..."** Rupert Read (2002), p. 208.

281   **"We cannot say what time is ..."** Lucas (1973), p. 4.

281   **"because we instinctively ..."** Lockwood (2005), p. 53.

282   **"From this perspective ..."** ibid., pp. 53-54.

282   **"I would say yes ..."** Barbour interview (2007).

284   **... proposed alternative to string theory ...** For Smolin's own account of the theory, see Lee Smolin, "Atoms of Space and Time," *Scientific American,* Jan. 2004, pp. 66-75.

284   **"... involve small loops ..."** ibid., p. 69.

284   **"A one-sentence comparison ..."** Greene (2004), p. 489.

284   **... has promised the world ...** Smolin (2006).

285   **"philosophical guru"** Smolin (1997), p. 223.

285   **"sloppy thinking"** Lee Smolin interview, December 1997.

286   **"If one takes the positivist position ..."** Hawking (2001), p. 31.

287   **"Speaking personally ..."** Smolin (1997), p. 286.

288   **"What we experience ..."** Deutsch (1997), p. 263.

288   **"If I show you ..."** David Deutsch interview, May 2007.

290   **"Does this road go anywhere?"** D.C. Williams, "The Myth of Passage," in Westphal and Levenson (1994), p. 137.

290   **"I do not believe ..."** quoted in Yourgrau (2005), p. 111.

290   **"The objective world simply ..."** quoted in Gell (1992), p. 154.

290   **"We experience the world in time ..."** Lee Smolin, presentation to the New York Academy of Sciences, Oct. 15, 2007.

291   **"... a way to *unfreeze* time ..."** Smolin (2006), p. 257.

291   **"... deeply wrong about the nature of time"** Smolin (2006), p. 256.

291   **"I wish time were ..."** quoted in Dennis Overbye, "On Gravity, Oreos, and a Theory of Everything," *The New York Times,* Nov. 1, 2005 (online edition).

291   **"As a physicist ..."** Davies (1995), p. 275.

292   **"not yet hanged"** Kaku (2004), pp. 178-179; see also Fölsing (1997), p. 666 and Isaacson (2007), p. 304.

294   **"Should I still be wondering ..."** Yourgrau (2005), p. 123.

295   **"Once Einstein said that ..."** Rudolf Carnap, "Intellectual Autobiography," in P.A. Schilpp (ed.), *The Philosophy of Rudolph Carnap* [sic]. La Salle, Ill.: Open Court, 1963. pp. 37-38.

296   **"Now he has departed ..."** quoted in Calaprice (2005), p. 73.

BIBLIOGRAPHY

Titles marked with a ★ are especially recommended for their treatment of particular aspects of the nature of time. Those marked ★ (T) are also recommended but contain some technical material that may be suitable for those with a background in the physical sciences.

★ Adams, Fred and Gregory Laughlin. *The Five Ages of the Universe.* New York: The Free Press, 1999.

Alexander, H.G. *The Leibniz-Clarke Correspondence.* Manchester: Manchester University Press, 1956.

Aveni, Anthony. *Ancient Astronomers.* Washington: Smithsonian Books, 1995.

★ ____. Anthony. *Empires of Time.* New York: Kodanasha International, 1995.

Barnes, Jonathan. *Early Greek Philosophy.* London: Penguin, 1997.

Benford, Gregory. *Deep Time.* New York: HarperCollins, 1999.

★ Blaise, Clark. *Time Lord: Sir Sandford Fleming and the Creation of Standard Time.* London: Weidenfeld & Nicholson, 2000 (2001 ed.).

Bostrom, Nick, and Milan Cirkovic (eds.). *Global Catastrophic Risk.* Oxford: Oxford University Press, in press.

Brandon, S.G.F. *History, Time and Deity.* Manchester: Manchester University Press, 1965.

Brockman, John (ed.). *How Things Are: A Science Tool-kit for the Mind.* London: Weidenfeld & Nicholson, 1995.

★ Boorstin, Daniel. *The Discoverers: A History of Man's Search to Know His World and Himself.* New York: Random House, 1983 (1985 ed.).

Burl, Aubrey. *The Stone Circles of the British Isles.* New Haven, Conn.: Yale University Press, 1976.

Calaprice, Alice (ed.). *The New Quotable Einstein,* Princeton, Princeton University Press, 2005.

Clute, John, and Peter Nicholls (ed). *The Encyclopedia of Science Fiction.* New York: St. Martin's Press, 1995.

Cocks, Doug. *Deep Futures: Our Prospects for Survival.* Montreal and Kingston: McGill-Queen's University Press, 2003.

Cohen, I. Bernard, and Anne Whitman. *Isaac Newton – The Principia: A New Translation*. Berkeley: University of California Press, 1999.

Cornish, Edward. *The Study of the Future*. Washington: World Future Society, 1977.

Coveney, Peter, and Roger Highfield. *The Arrow of Time: A Voyage through Science to Solve Time's Greatest Mystery*. New York: Ballantine Books, 1990.

★(T) Dainton, Barry. *Time and Space*. London: Acumen Publishing, 2001.

★ Danielson, Dennis (ed.). *The Book of the Cosmos: Imagining the Universe from Heraclitus to Hawking*. Cambridge, Mass: Perseus Publishing, 2000.

Dale, Rodney. *Timekeeping*. London: The British Library, 1992.

★ Davies, Paul. *About Time*. London: Penguin Books, 1995.

_____. *How to Build a Time Machine*. London: Penguin Books, 2001.

★ Deutsch, David. *The Fabric of Reality*. London: Penguin Books, 1997.

★ Duncan, David Ewing. *Calendar: Humanity's Epic Struggle to Determine a True and Accurate Year*. New York: Avon Books, 1998.

Einstein, Albert (trans./ed. Paul A. Schilpp). *Autobiographical Notes*. Chicago: Open Court Publishing, 1979.

Einstein, Albert, and Leopold Infeld. *The Evolution of Physics*. New York: Simon and Schuster, 1938 (1966 ed.).

Falk, Dan. *Universe on a T-Shirt: The Quest for the Theory of Everything*. Toronto: Penguin Books, 2002.

★ Ferris, Timothy. *Coming of Age in the Milky Way*. New York: Anchor Books, 1988 (1989 ed.).

★ _____. *The Whole Shebang*. New York: Simon & Schuster, 1997 (1998 ed.).

★ _____ (ed.). *The World Treasure of Physics, Astronomy, and Mathematics*. New York: Little, Brown and Company, 1991.

★ Fölsing, Albrecht. *Albert Einstein*. New York: Penguin Books, 1997 (1998 ed.).

★ Fraser, J.T. *Time: The Familiar Stranger*. London: Tempus Books, 1987.

Galilei, Galileo. *Dialogue Concerning the Two Chief World Systems – Ptolemaic and Copernican* (trans. Stillman Drake). Berkeley: University of California Press, 1967.

Galison, Peter. *Einstein's Clocks, Poincaré's Maps*. New York: W.W. Norton & Company, 2003.

★ Gell, Alfred. *The Anthropology of Time: Cultural Constructions of Temporal Maps and Images*. Oxford: Berg, 1992.

★ Gleick, James. *Isaac Newton*. New York: Random House, 2003 (2004 ed.).

★ Gorst, Martin. *Measuring Eternity*. New York: Broadway Books, 2001.

Gott, J. Richard. *Time Travel in Einstein's Universe: The Physical Possibilities of Travel through Time*. New York: Houghton Mifflin, 2002.

★ (T) Greene, Brian. *The Elegant Universe.* New York: W.W. Norton & Company, 1999.

★ (T) _____. *The Fabric of the Cosmos.* New York: Vintage Books, 2004.

Gribbin, John. *In Search of Schrödinger's Cat: Quantum Physics and Reality.* New York: Bantam Books, 1984 (1988 ed.).

★ _____. *The Birth of Time: How Astronomers Measured the Age of the Universe.* New Haven: Yale University Press, 1999.

_____. *The Origins of the Future: Ten Questions for the Next Ten Years.* New Haven: Yale University press, 2006.

Griffiths, Sian (ed.). *Predictions.* Oxford: Oxford University Press, 1999.

Hawking, Stephen. *Black Holes and Baby Universes.* New York: Bantam Books, 1994.

_____. *A Brief History of Time.* New York: Bantam Books, 1988.

★ _____. *The Universe in a Nutshell.* New York: Bantam Books, 2001.

★ Isaacson, Walter. *Einstein: His Life and Universe.* New York: Simon & Schuster, 2007.

★ Kaku, Michio. *Einstein's Cosmos: How Albert Einstein's Vision Transformed Our Understanding of Space and Time.* New York: W.W. Norton & Company, 2004.

_____. *Visions: How Science Will Revolutionize the 21st Century.* New York: Anchor Books, 1997.

Kandel, Eric. *In Search of Memory.* New York: W.W. Norton & Company, 2006.

★ Klein, Richard G., with Blake Edgar. *The Dawn of Human Culture.* New York: John Wiley & Sons, 2002.

Krauss, Lawrence. *The Physics of Star Trek.* New York: Basic Books, 1995.

Kurzweil, Ray. *The Age of Spiritual Machines.* New York: Penguin Books, 1999 (2000 ed.).

★ Landes, David S. *Revolution in Time: Clocks and the Making of the Modern World.* Cambridge, Mass.: Harvard University Press, 1983.

★ Levine, Robert. *A Geography of Time: The Temporal Misadventures of a Social Psychologist, or How Every Culture Keeps Time Just a Little Bit Differently.* New York: Harper Collins, 1997.

Leslie, John. *The End of the World: The Science and Ethics of Human Extinction.* New York: Routledge, 1996.

★ Lippincott, Kristen (ed). *The Story of Time.* London: Merrell Holberton Publishers, 2000.

★ (T) Lockwood, Michael. *The Labyrinth of Time,* Oxford: Oxford University Press, 2005.

Lucas, J. R. *A Treatise on Time and Space.* London: Methuen & Co. Ltd., 1973.

Macey, Samuel L. (ed.). *The Encyclopedia of Time.* New York: Garland Publishing, 1994.

Mallett, Ronald, with Bruce Henderson. *The Time Traveler.* New York: Thunder's Mouth Press, 2006.

McCready, Stuart (ed.). *The Discovery of Time.* Naperville, Ill.: Sourcebooks Inc., 2001.

★ Mithen, Steven. *The Prehistory of the Mind.* London: Thames and Hudson, 1996.

★ Nahin, Paul. *Time Machines: Time Travel in Physics, Metaphysics, and Science Fiction.* New York: Springer Verlag, 1999.

★(T) Pais, Abraham. *Subtle is the Lord: The Science and Life of Albert Einstein.* Oxford: Oxford University Press, 1982.

★(T) Penrose, Roger. *The Emperor's New Mind.* New York: Oxford University Press, 1989 (1990 ed.).

_____. *The Road to Reality: A Complete Guide to the Laws of the Universe.* New York: Alfred A. Knopf, 2005.

_____. *Shadows of the Mind.* Oxford: Oxford University Press, 1994 (1995 ed.).

Pickover, Clifford A. Time: *A Traveler's Guide.* Oxford: Oxford University Press, 1998.

Price, Huw. *Time's Arrow and Archimedes' Point: New Direction for the Physics of Time.* Oxford: Oxford University Press, 1996.

Pritchard, Evan T. *No Word for Time: The Way of the Algonquin People.* Tulsa, Okla.: Council Oak Books, 1997.

Rees, Martin. *Our Cosmic Habitat.* Princeton: Princeton University Press, 2001.

★ _____. *Our Final Hour.* New York: Basic Books, 2003.

Ridderbos, Katinka (ed). *Time.* Cambridge: Cambridge University Press, 2002.

Ruggles, Clive. *Astronomy in Prehistoric Britain and Ireland.* New Haven, Conn.: Yale University Press, 1999.

Savitt, Steven (ed). *Time's Arrow Today: Recent philosophical work on the direction of time.* Cambridge: Cambridge University Press, 1995.

★ Schacter, Daniel. *Searching for Memory.* New York: Basic Books, 1996.

★ _____. *The Seven Sins of Memory.* New York: Houghton Mifflin Company, 2001.

Shapley, Harlow et. al. (eds.). *A Treasury of Science.* New York: Harper and Brothers, 1943.

Smolin, Lee. *The Life of the Cosmos.* Oxford: Oxford University Press, 1997.

★ _____. *The Trouble With Physics: The Rise of String Theory, the Fall of a Science, and What Comes Next.* New York: Houghton Mifflin Company, 2006.

★ Sobel, Dava. *Longitude: The True Story of a Lone Genius Who Solved the Greatest Scientific Problem of His Time.* New York: Penguin Books, 1995 (1996 ed.).

Stachel, John. *Einstein's Miraculous Year.* Princeton: Princeton University Press, 1998 (2005 ed.).

★ Steel, Duncan. *Marking Time: The Epic Quest to Invent the Perfect Calendar.* New York: John Wiley & Sons, 2000.

★ (T) Thorne, Kip. *Black Holes and Time Warps*. New York: W.W. Norton & Company, 1994.

Toomey, David. *The New Time Travelers*. New York: W.W. Norton & Company, 2007.

★ Toulmin, Stephen, and June Goodfield. *The Discovery of Time*. Chicago: University of Chicago Press, 1965 (1977 ed.).

Turetzky, Philip. *Time*. London: Routledge, 1998.

★ Weinberg, Steven. *The First Three Minutes*. New York: Basic Books, 1997 (1988 ed.).

★ Westfall, Richard. *The Life of Isaac Newton*. Cambridge: Cambridge University Press, 1994.

Westphal, Carl, and Jonathan Levenson (eds.). *Reality*. Indianapolis: Hackett Publishing Co., 1994 (1993 ed.).

Whitrow, G. J. *The Nature of Time*. London: Penguin, 1972 (1975 ed.).

★ ____. *Time in History: Views of Time from Prehistory to the Present Day*. Oxford: Oxford University Press, 1988 (1990 ed.).

Yourgrau, Palle. *A World Without Time: The Forgotten Legacy of Gödel and Einstein*. New York: Basic Books, 2005 (2006 ed.).

INDEX

DAN FALK has written about science for the *Globe and Mail*, the *Toronto Star*, *The Walrus*, the *New Scientist*, and *Nature*, among other publications, and has been a regular contributor to the CBC Radio program *Ideas*. His awards include gold and silver medals for his radio documentary work from the New York Festivals and the Science Writing Award in Physics and Astronomy from the American Institute of Physics. His first book, *Universe on a T-Shirt*, won the 2002 Science in Society Journalism Award from the Canadian Science Writers' Association. He lives in Toronto.

Please visit his website at www.danfalk.ca.